Robert M. Augros & George N. Stanciu

The New Story of Science

Mind and the Universe

Preface by Sir John Eccles

D1310233

A Gateway Edition
Regnery Gateway

The authors would like to thank the Director,
Montreal Neurological Institute and Princeton
University Press, Dover Publications, and Editions
D'Art Albert Skira, S. A. for their Kind permission
to reprint the illustrations in this book.

Published by Regnery Gateway, Inc.
940-950 North Shore Drive
Lake Bluff, Illinois 60044

Library of Congress Cataloging in Publication Data

Augros, Robert M., 1943–
    The new story of science.

    Bibliography: p.
    Includes index.
    1. Science—Philosophy. 2. Intellect. 3. Man.
4. Cosmology. I. Stanciu, George N., 1937–
II. Title.
Q175.A874   1984   501   84-43228
ISBN 0-89526-833-7 (pbk.)

CONTENTS

PREFACE
by
Sir John Eccles

It is a great honour and responsibility to be writing a preface to this most unusual book. We are told that the concept of "New Story" comes from a recent article by the cultural historian Thomas Berry. By a "Story" is meant a "cosmic world view of a civilization, the universal framework in which all else is understood and evaluated." The thesis of this book is that there was in the 17th, 18th and 19th centuries the gradual building up of a world view in physics and cosmology that was progressively more materialistic and which the authors refer to as the "Old Story." The great explanatory power of physical science with its immense display of technology has deluded the "common man" so that he is less and less oriented to religious belief and spiritual values. The concepts of mind and mental events have been threatened and even rejected. However, now in the 20th century the revolution in physics and cosmology has transformed the story because in quantum physics the conscious observer has become essential as a participator in the scientific measurements; and the origin of the universe in a unique cosmic explosion, the Big Bang, about 12 thousand million years ago alters the concepts of space and time and ushers in the wonderful story of contingencies that leads from our solar system to planet Earth, to the origin of life, to biological evolution and so to human persons. From being an insignificant creature on a modest planet circling around an ordinary star in an immense galaxy of 100 thousand million other stars, man has now become exalted as a participator in a great cosmic drama, there being even the concept of the Anthropic Principle, that the whole cosmic happening from the Big Bang onwards was designed in order that conscious observers could exist at some place and time in the expanding Universe. This is convincing enough evidence that there is coming about a new cosmic world view, the Old Story giving place to a New Story which is built around human persons as conscious observers and participators, and in which mind and mental events have a status matching that of the material world.

There is dramatic simplicity in the global concepts of "old story" and "new story," and the authors are to be commended for daring to present the on-going conflict in terms that the general reader can understand. Actually there is much confusion in the tangle of conflicting ideologies. I can understand the "old story" and "new story" dramatic device quite clearly for physics and cosmology. In my own field of expertise, the mind-brain problem, I would hesitate to define sharply the old story from the new story. Scientifically I would still adhere to most of the old story, but would deviate philosophically. We are caught in between the old story and the birth-throes of the new story. And this is even true of physics, where Newtonian physics survives as a necessary basis of all technology, even the return journey to the moon.

It is the merit of this book that it enables us to assess the on-going conflicts between the old story and the new story. It gives us a vantage point for watching the battle lines with their gains and losses. But there is the risk that the conflict in ideologies may be too starkly drawn so that some readers would not feel committed. Furthermore there are large areas of actual conflict not dealt with by the authors. For example evolution and genetics, sociobiology and anthropology, the biological evolution of animal consciousness and the miraculous origin of human self-consciousness. When dealing with the new dualist world view the philosophy of Karl Popper with Worlds 1, 2, and 3 is missing. Also I find the treatment of artistic creativity unconvincing. So there is much that could be dealt with in a second volume on the same general theme.*

I am sure that this book will arouse controversy from the old story devotees. Most will not be convinced, but the great appeal will be to the young who have been outraged by arrogant statements such as those of Russell quoted in the introduction. We all are repelled by the ideology of a meaningless existence with unyielding despair. The appeal of the New Story is that it replaces this terrible starkness with "purpose, God, beauty, the spiritual goods, and the dignity of man," as stated in the final conclusion of the book.

John C. Eccles
Locarno, Switzerland

*The authors are presently working on a book on the life sciences and plan another volume on the artist.

INTRODUCTION

Every civilization has a cosmic world view—a Story according to which all else is understood and evaluated. The prevailing Story shapes a culture's attitudes, integrates its knowledge, dictates its methodology, and directs its education. The Story acts as the context and the measure of further knowledge. A world view is so fundamental that we realize we have one only when confronted by an alternative— either through travelling to another culture, or by reading of past ages, or when the world view of our own culture is under transition.

Since the Renaissance, Western culture has been under the sway of experimental science. But the world view conceived during the Renaissance is currently being challenged by twentieth century science. This gives rise to two contending Stories of science. As cultural historian Thomas Berry puts it: "It's all a question of story. We are in trouble just now because we do not have a good story. We are between stories. The Old Story . . . is not functioning properly, and we have not learned the New Story."[1]

The Old Story of science is scientific materialism. It holds that only matter exists and that all things are explicable in terms of matter alone. Thus, free choice must be an illusion, since matter cannot act freely. And since matter cannot plan or aim at anything, purpose cannot be found in natural things. Mind itself is considered to be a by-product of brain activity. Bertrand Russell, with characteristic force and clarity, portrays the place of man in the Old Story:

"That man is the product of causes which had no provision of the end they were achieving; that his origins, his growth, his hopes and fears, his loves and beliefs are but the outcome of accidental collocations of atoms; that no fire, no heroism, no intensity of thought and feeling, can preserve an individual life beyond the grave; that all the labors of the ages, all the devotion . . . all the noonday brightness of human genius are destined to extinction . . . all these things, if not quite beyond dispute, are yet so nearly certain, that no philosophy which rejects them can hope to stand. Only within the scaffolding of these truths, only on the firm foundation of unyielding despair, can the soul's habitation henceforth be safety built."[2]

But since 1903 when Russell wrote those words, science has undergone a series of dramatic revolutions: first in physics with Einstein, Bohr, and Heisenberg; then in neuroscience with Sherrington, Eccles, Sperry, and Penfield; in psychology with Frankl, Maslow, and May; and in cosmology with the Big Bang and the Anthropic Principle. These discoveries have not only transformed the modern conception of man and his place in the world; they quite unexpectedly tell a New Story, rather different from Russell's account. Physicist Henry Margenau remarks, "The basic tenet of materialism is that all reality consists of matter, a view fairly tolerable at the end of the last century. . . . But much has happened in the meantime that discredits this view." Physicist Werner Heisenberg declares that modern "atomic physics has turned science away from the materialistic trend it had in the nineteenth century."[3]

A world view is so fundamental that it cannot be changed easily or quickly—even if there is compelling evidence for a change. Instead, there is always a tendency to force the new knowledge into the old view. As molecular biophysicist Harold Morowitz remarks:

"Really deep concepts seem to take about 50 years to sink into the collective conscience of the thinking community. So it is that only now are most of us beginning to sense the full impact of certain ideas that have been brewing in physics since the first quarter of this century."[4] The present book is an attempt to put together the elements of this New Story of science.

In using the metaphors Old Story of science and New Story of science we do not mean to imply that all scientists are divided neatly into two camps. The two Stories more often show themselves as general attitudes of mind, and not everyone embraces all the consequences of either Story. Nor did an Old Story view of the sort just seen in Russell appear fully formed when modern science first began. There has been a certain historical development as we shall see. Furthermore, in criticizing the Old Story we in no way desire to call into question the wealth of truths it uncovered. Finally, we intend to contrast the Old Story and the New Story not in their technical details, but as world views with the focus on the broad, philosophical implications for man and the universe.

I

MATTER

A revolution overthrows an old regime and initiates a new order. Therefore, to comprehend the significance of any revolution it is necessary first to understand the old regime. In modern physics the old regime is the Newtonian system. The essence of this system can be seen in the way it answers three questions.

What are the constituents of the physical universe? There are three realities: matter, space, and time. Matter, according to Newton, is composed of "massy, hard, impenetrable, moveable particles, of [various] sizes and figures." As the properties of matter Newton lists "extension, hardness, impenetrability and inertia." The nature and the properties of these particles, the atoms, are fixed for all time. The atom is considered to be the ultimate particle.[1]

Both time and space are conceived to be absolutes; that is, they would continue to exist even if all material things in the universe were annihilated. Newton describes space in these terms: "Absolute space, in its own nature, without relation to anything external, remains always similar and immovable." He adds, "Absolute, true, and mathematical time, of itself, and from its own nature, flows equally without relation to anything external." Both time and space are thought to be infinite in extent, universal, and unchangeable.[2]

What is change? Newton explains: "The changes of corporeal things are to be placed only in the various separations and new associations and motions of these permanent particles."[3]

How do changes occur? Physical laws govern the motion of matter in absolute space and time. Newton describes the ideal of his system: "To derive two or three general principles of motion from phenomena, and afterwards to show how the properties and actions of all corporeal things follow from those manifest principles, would be a great step in philosophy."[4] The scientist has no place in this system except as a detached spectator. It was assumed that the physical universe and all the properties of matter could be understood without introducing mind into the system.

The Newtonian system met with many

successes, especially in physics and chemistry. Carried forward by Faraday, Kelvin, Herschel, and hundreds more, the old regime triumphed in explaining phenomena of motion, heat, light, and electricity. Understandably, these successes inspired a wish to extend this mode of explanation to all disciplines, including biology, psychology, history, and economics. The possibility of explaining so much of the natural world by assuming matter alone gradually led some to consider materialism as a part of the scientific method itself. The scientist, regardless of his private beliefs, would proceed in his scientific arguments on the presumption that matter alone is real, or at least that matter is all we can know scientifically. We might call this "methodological materialism." Newton himself was clearly not a materialist. He hoped by his mechanics to explain not *all* things but "all *corporeal* things."

The program endured and held out great promise. The scientists of the nineteenth century had every reason to expect that the twentieth century would bring the completion of the system. Many physicists thought their part of the account was already essentially completed.

The new century did bring major breakthroughs, but not at all of the sort expected. The new discoveries did not complete Newtonian physics; they overthrew

it. First, in 1905, Albert Einstein overturned two main pillars of the old regime:
special relativity led physics to abandon
forever the notions of absolute space and
absolute time. Einstein showed that
space-time and the laws of motion can be
defined only by reference to an observer
and his physical conditions. Other features
of special relativity theory such as the
equivalence of matter and energy are in
fact consequences of this centrality of the
observer. With special relativity the observer suddenly became an essential part
of the world of physics. The scientist could
no longer consider himself a detached
spectator as in the Newtonian system.

Next, a similar revolution occurred in
particle physics. In 1911, Ernest Ruther-
ford demonstrated that the atom consists
of an extremely small nucleus surrounded
by a swarm of electrons. Physicists at-
tempted to explain the structure of the
atom on the basis of Newtonian physics,
but every such attempt led to frustrating
contradictions. These failures finally resulted in the complete abandonment of the
Newtonian system at the atomic level and
precipitated the development of quantum
mechanics in the 1920's by such men as
Niels Bohr and Werner Heisenberg. With
the advent of quantum mechanics, the role
of the observer became even more crucial
in physical theory. Physicist Max Born

writes, "No description of any natural phenomenon in the atomic domain is possible without referring to the observer, not only to his velocity as in relativity, but to all his activities in performing the observation, setting up the instruments, and so on."[5] Physicist John Wheeler explains:

"It was long natural to regard the observer as in effect looking at and being protected from contact with existence by a 10 cm. slab of plate glass . . . In contrast, quantum mechanics teaches the direct opposite. It is impossible to observe even so miniscule an object as an electron without in effect smashing that slab and reaching in with the appropriate measuring equipment . . . Moreover, the installation of apparatus to measure the position coordinate, x, of the electron automatically prevents the insertion in the same region at the same time of the equipment that would be required to measure its velocity or its momentum, p; and conversely. The act of measurement typically produces an unpredictable change in the state of the electron. This change is different according as one measures the position or the momentum. The choice one makes about what one observes makes an irretrievable difference in what one finds. The observer is elevated from 'observer' to 'participator.' What philosophy suggested in times past, the central feature of quantum mechanics tells us today with impressive force. In some strange sense this is a participatory universe."[6]

Thus the properties of the smallest particles of matter cannot be defined independently of the choices and actions of the observer who is necessary not only as witness but as participant. Physicist Eugene Wigner points out the implications for the role of the mind in the world: "When the province of physical theory was extended to encompass microscopic phenomena, through the creation of quantum mechanics, the concept of consciousness came to the fore again: it was not possible to formulate the laws of quantum mechanics in a fully consistent way without reference to the consciousness." And since matter at the most basic level is understood only by introducing mind, Wigner concludes that mind is one of nature's ultimate realities: "There are two kinds of reality or existence: the existence of my consciousness and the reality or existence of everything else. . . . It is profoundly baffling that the existence of the first kind of reality could ever be forgotten."[7]

Wigner describes the outlook of the old regime:

"Until not many years ago, the 'existence' of a mind or soul would have been passionately denied by most physical scientists. The brilliant successes of mechanistic and, more generally, macroscopic physics and of chemistry overshadowed the obvious fact that thoughts, desires, and emotions are not made of matter, and it was nearly universally accepted among physical scientists that there is nothing besides matter."[8]

Relativity and quantum mechanics, then, exhibit a common departure from the Newtonian Story: the introduction of mind. Physics in the twentieth century has gradually replaced materialism with the affirmation that mind plays an essential role in the universe. It is most remarkable for such an affirmation to come out of the science of physics. If materialism succeeds anywhere one would expect it to succeed in the study of matter itself.

The new facts of relativity and quantum mechanics cannot be fitted into the old view. Neither the structure of space-time nor the properties of elementary particles can be described without reference to an observer-participant, a mind. The Old Story contained only matter and physical laws; the New Story of science must contain matter, physical laws, and *mind*.

II

MIND

The principle of simplicity seems to require that science explain natural things in terms of matter alone unless such a procedure proves to be unworkable. In this context the Old Story's approach to mind makes a good deal of sense. It argues that all the properties of natural things ultimately arise from interactions between the particles of which they are composed. Thus, water is so fluid because water molecules slide by each other with little friction; rubber is elastic because rubber molecules are easily deformed but resilient; a diamond is extremely hard because the carbon atoms in it are tightly bound in a rigid lattice structure. The same must hold for the mind. Nineteenth century biologist Thomas H. Huxley states: "The thoughts to which I am giving utterance, and your thoughts regarding them, are the expression of molecular changes."[1] In the Old Story's perspective, the best way to investigate the mind is to show how the mind is derived from matter.

One implication of this view is that the human mind cannot choose freely since matter acts only by mechanical necessity. This is why there is an inclination in the Old Story to explain human behavior in terms of the instinct, physiology, chemistry, and physics. There is no room for free choice. In fact, if one assumes a rigorous materialism, any influence of the mind or will on the brain must be denied. Material changes cause thought, not the reverse. Nineteenth century mathematician W.K. Clifford in a lecture on science puts it succinctly: "If anyone says that the will influences matter, the statement is not untrue, it is nonsense."[2] Huxley characterizes the relation of mind and body in this way:

"Consciousness . . . would appear to be related to the mechanisms of the body . . . simply as a side product of its working and to be completely without any power of modifying that working as the sound of a steam whistle which accompanies the work of a locomotive . . . is without influence upon its machinery."[3]

A second implication of the Old Story's

conception of mind is that nothing in man can survive death. If thinking and willing are activities of the brain, then there is no reason to suppose those activities can continue after the brain is destroyed. If every part of man is matter, then every part of man is mortal. Only matter is eternal in the Old Story.

The Old Story, then, had a clear program to account for mind. But no one in the nineteenth century could say precisely how mind arises from matter. Physiologists expected the future to bring the answer. In 1868, Huxley wrote, "so will the physiology of the future gradually extend the realm of matter and law until it is co-extensive with knowledge, with feeling, and with action."[4] Many looked to the twentieth century for the fulfillment of the materialistic program.

The new century did bring extraordinary discoveries in physiology, but not at all of the sort expected. The new findings did not complete the Old Story; they advanced a New Story.

It began with Sir Charles Sherrington, considered the founder of modern neurophysiology. As a result of his pioneering investigations of the nervous system and the brain, Sherrington concluded, "A radical distinction has therefore arisen between life and mind. The former is an affair of chemistry and physics; the

latter escapes chemistry and physics."[5] By
life, Sherrington refers to self-nutrition,
cell metabolism, and growth. Such
phenomena, he says, are accomplished by
means of the laws of physics and chemistry
and can be explained in terms of them. The
actions of the mind, however, transcend
the mechanisms of physics and chemistry.

Neuroscientist Sir John Eccles agrees:
"Conscious experiences . . . are quite different in kind from any goings-on in the
neuronal machinery; nevertheless the
events in the neuronal machinery are a
necessary condition for experience, though
. . . they are not a sufficient condition."[6]

Let us take an example to illustrate
what Eccles and Sherrington are saying.
What happens when Socrates sees a tree,
for instance? Sunlight reflecting from the
tree enters the pupil of Socrates' eye and
passes through the lens which focuses a
miniature inverted image of the tree on
the retina, causing there physical and
chemical changes. Is this seeing? No. If
Socrates were unconscious, the same
image could be focused on his retina, causing the same physical and chemical
changes, but in such a case he would perceive nothing. Likewise, a camera focuses
an image and its film undergoes physical
and chemical changes, but the camera does
not literally see the colors and forms that it
records.

To explain Socrates' seeing we need much more. The retina, a sheet of closely-packed receptors (ten million cones and one hundred million rods), when activated by the light from the tree, begins firing impulses to the optic nerve which transmits them to the visual cortex of the brain. So far everything can be explained by physics and chemistry. But where does *green* come in? The brain itself is grey and white. How can it receive a new color without losing its old color? And how does Socrates' brain perceive light if his brain is completely enclosed and shut off from all light? It would be understandable if, when Socrates directed his eyes to the tree, all he experienced was a whir of electricity in his own brain. But the electrical and chemical activity of his brain, that somehow makes possible his seeing, is precisely what Socrates does not perceive. Instead, when he sees, Socrates perceives colors, shapes, movements, and light—all in three dimensions. It is difficult even to imagine how any of these might arise out of chemicals and electricity.

Eccles underscores the mystery in sense perception:

"Is it not true that the most common of our experiences are accepted without any appreciation of their tremendous mystery? Are we not still like children in our outlook on our experiences of conscious life, accepting them and only rarely pausing to contemplate and appreciate the wonder of conscious experience? For example, vision gives us from instant to instant a three-dimensional picture of an external world and builds into that picture such qualities as brightness and color, which exist only in perceptions developed as a consequence of brain action. Of course, we now recognize physical counterparts of these perceptual experiences, such as the intensity of the radiating source and the wave lengths of its emitted radiation; nevertheless, the perceptions themselves arise in some quite unknown manner out of the coded information conveyed from the retina to the brain."[7]

The image focused on the retina, for instance, never reappears as such in the brain. The conscious mind must reconstitute it from coded impulse patterns. Every sense perception consists of three stages: the original stimulus to the sense organ, the nerve impulses sent to the brain, and the pattern of neuronal activity evoked in the brain. Eccles summarizes:

"The transmission from sense organ to cerebral cortex utilizes a coded pattern of nerve impulses that may be likened to a Morse Code with dots only in various temporal sequences. Certainly, this coded transmission is quite unlike the original stimulus to that sense organ, and the spatio-temporal pattern of neuronal activity that is evoked in the cerebral cortex . . . is again quite different."[8]

This double translation amplifies the wonder of sense perception. That such a chain of physical-chemical translations should result in the particular sensory experience "green" is no less startling than a person suddenly understanding a text when it is translated from one language unknown to him into another language unknown to him!

According to the New Story, then, the world of sensation depends on the world of physics and chemistry but is not reducible to it. A comparison may serve to clarify this subtle dependence: a book certainly depends on the paper, the glue, and the ink which compose it. Without them it could not exist. And yet the book is not adequately understood by a mere chemical analysis of the ink and paper fibers. Even an exhaustive knowledge of every molecule of paper and ink would yield nothing about the book's content. The content of a book comprises a higher order that transcends the world of physics and chemistry. In a similar way, the New Story affirms that our sensations depend on bodily organs but cannot be reduced to the physical and chemical properties of matter.

Using the example of sight, Sherrington explains that the Old Story, which he calls the "energy-scheme," cannot account for the sensation of

"a star which we perceive. The energy-scheme deals with it, describes the passing of radiation thence into the eye, the little light-image of it formed at the bottom of the eye, the ensuing photo-chemical action in the retina, the trains of action-potentials travelling along the nerve to the brain, the further electrical disturbance in the brain . . . But, as to our *seeing* the star it says nothing. That to our perception it is bright, has direction, has distance, that the image at the bottom of the eye-ball turns into a star overhead, a star moreover that does not move though we and our eyes as we move carry the image with us, and finally that it is the thing a star, endorsed by our cognition, about all this the energy-scheme has nothing to report. The energy-scheme deals with the star as one of the objects observable by us; as to the perceiving of it by the mind the scheme puts its finger to its lip and is silent. It may be said to bring us to the threshold of the act of perceiving, and there to bid us 'good-bye.' Its scheme seems to carry us to and through the very place and time which correlate with the mental experience, but to do so without one hint further."[9]

In the New Story of science, then, the physiological and chemical activity of the brain are necessary for sensation and are simultaneous with it but they are not the same as the sensation itself. Matter alone cannot account for sense perception. The Old Story can speak of light waves, of chemical changes, of electrical impulses in nerves, and of brain cell activity. But as to seeing, smelling, tasting, hearing, and touching themselves, materialism has nothing to say.

Sense perception is real yet it is not matter, it is not a quality of matter, and it cannot be explained by matter. Hence Sherrington concludes, "That our being should consist of two fundamental elements offers, I suppose, no greater inherent improbability than that it should rest on one only."[10] The New Story posits two irreducible elements in man: body and mind.

So far we have examined only the example of sense perception. What does the New Story have to say about the human intellect? Before we broach that question, however, it will be necessary first to distinguish the intellect clearly from our other mental faculties. We shall do this briefly and in a common sense fashion, relying on the internal experience that all of us share.

The word *conscious* means "with knowledge or awareness." In this sense, any activity that implies knowledge or awareness is an activity of the conscious mind. It all begins with sense perception. The external senses are the first foundation and source of all human knowledge. Without the information coming from them, the memory would have nothing to remember, the imagination nothing to imagine, the intellect nothing to understand. Each of the five external senses—sight, smell, hearing, taste, and touch—perceives a specific quality of physical things. Sight alone perceives colors; hearing, sounds; smell, odors; taste, flavors. The sense of touch perceives temperatures, textures, and pressures. Certain other qualities such as size and shape are perceivable by more than one sense. For example, we can tell the size of a coin by either sight or touch. Of the five external senses, only touch is distributed throughout the body. The other four are limited each to a specialized organ: the eye, the ear, the nose, the tongue.

In addition to the external senses, we have an array of internal sense powers at our disposal. Consider for a moment that we are able to sense not only white and sweet, but we are able to perceive also the difference between the two. The eye perceives white but not sweet. The tongue perceives sweet but not white. Neither the tongue nor the eye can discern white from sweet since neither perceives both. The same holds for the difference between loud and hot. Any faculty that compares two things must know both. No external sense can perform this task. Therefore, there must be in us an internal sense that can perceive and distinguish all the qualities apprehended by the external senses.

We also have the capacity to call to mind something no longer present. The activity of remembering is something actually present, but the thing remembered is not. Somehow our original sense perception passed away and yet is still available to us. Memory not only recalls the previous experience but recalls it *as past,* and can catalogue it chronologically in relation to other past experiences. Even more amazing is our ability to bring ourselves to recall *something forgotten.* Memory sometimes fails as when we cannot remember a person's name. But we can often bring ourselves to recall it by concentrating on other things that are associated with the name.

Imagination is another interior sense power by which we can picture not only things perceived by the five external senses but even things not perceived by them such as a gold mountain or a flea-sized elephant. The imagination, unlike the memory, works freely and creatively on the information coming from the five external senses.

Our capacity for emotion, that is, for love, anger, joy, fear, hope, desire, and sorrow, connects us with the world in still a different way. Every emotion is caused by the act of a sense power, either an external sense, the imagination, or the memory. For example, anger is provoked by the perception of some injury or insult; fear by the imagination of some future, threatening evil; sorrow by the perception of present pain or by the memory of some past suffering. It is also part of every emotion to be felt. In fact, emotions are sometimes called *feelings*, because they are so closely allied to the sense of touch. Nevertheless, though sensation always causes and accompanies every emotion, emotions themselves are not acts of sense perception. Fear does not designate the mere perception of an object but rather an attitude or reaction to that object. An emotion is not the *seeing* but a response to what is seen, inclining us toward it or propelling us away from it.

The higher animals exercise most of the capacities mentioned so far. But if man is more than just an animal, there must be some special capacity distinguishing him from the rest. A look at the hierarchy of living things leads to the discovery of that capacity.

We notice that plants, through growth,

move themselves but do not *know* where they are going. A tree sends its roots down into the soil, but not because it is *aware* that water and nutrients are there. Animals, on the other hand, perceive through their senses where they are going but not *why*. For instance, a bird, by its power of sight, selects appropriate materials for its nest. But the bird does not build a nest because it *understands* that it is necessary for procreation. The bird's reactions are triggered automatically by certain stimuli. The warm spring sun causes the bird's pituitary gland to produce certain hormones that provoke nest-building activity. Birds injected with the female hormone estrogen will build nests out of season.

Plants move themselves but do not *know* where they are going. Animals perceive where they are going but not *why*. To complete the hierarchy there should exist creatures that know not only where they are going but also *why*. We ourselves are such creatures, and the power that enables us to understand the *why* of things is called mind or intellect. It is also called the power of reason, since by it we can discover the reasons for things. No sense power can perform that task. The tongue can tell us that the sea is salty but not why.

The mind also enables us to understand the *what* of things, which again, the senses cannot reach, not even the imagination. For example, if we try to imagine what an animal is, any picture we form in our sensory imaginations will be of this or that particular animal with a definite size, shape, and color. It is impossible to produce a sensory image of what all animals have in common. Yet it is not impossible for the mind to comprehend what an animal is.

Again, Einsteinian space cannot be pictured. Astrophysicist William Kaufmann writes, "It is virtually impossible to visualize a warped four-dimensional space-time."[11] Four-dimensional space cannot be sensed or imagined—even by the physicists and mathematicians—but it can be understood. In science the intellect rises above the restrictions of the imagination, an internal sense. The human mind, then, is not only distinct from the imagination, it is a greatly superior knowing power. The mind does science, not the senses. This is because only the mind can investigate the *what* and the *why* of things.

The mind is sometimes called the understanding, an apt name since the nature of a thing *stands under* the surface qualities apprehended by the senses. And the understanding also can penetrate to the cause that underlies or *stands under* the effect perceived by the senses. Hence, the name understanding is drawn from the mind's capacity to know the what and the why of things.

Finally, another faculty separates us from animals: the will. It is easy to distinguish the will from emotion since the two can conflict. Courageous actions prove that the will can assert itself even over the fear of death. The emotions are provoked by the senses, but the will chooses according to the judgment of reason. In fact, we often say a person overcame his emotion because he had good *reason* to do so. Animals follow the judgment of sensation and emotion, but man has the capacity to choose according to what his reason understands.

Now that we have seen what sets the human intellect and the human will apart from our other faculties, we can return to the question of what the New Story has to say about mind and will. Concerning the relation of brain and mind, some of the most remarkable discoveries of this century have been made during surgery on the brains of over a thousand conscious patients by Wilder Penfield. Regarding brain function, Penfield's observations surpass in authority and completeness all previous indirect evidence from research with animals and from brain surgery in anesthetized persons. Penfield, the man primarily responsible for integrating the disciplines of neurology, neurophysiology, and neurosurgery, began his pioneering research in the 1930's, but not until he published *The Mystery of the Mind* in 1975 did the full implications of his discoveries become evident.

Certain kinds of epilepsy are curable

through surgery. After administering general anesthesia and surgically removing a portion of the patient's skull to expose the brain, the surgeon revives the patient. Since the brain itself has no feeling, the surgeon can explore it with an electrode, and, with the patient's help, locate and remove the cells precipitating the epileptic seizures. (See Figure 1.)

Quite accidentally in 1933, Penfield found that gently stimulating certain regions of the brain with electricity triggers flashbacks of memory in a conscious patient. At first he was incredulous, but then he marveled: as his electrode touched the cortex of one young man, the man recalled being seated at a baseball game in a small town and watching a little boy crawl under the fence to join the audience. Another patient could hear instruments playing a melody. Penfield reports:

"I restimulated the same point thirty times trying to mislead her, and dictated each response to a stenographer. Each time I re-stimulated, she heard the melody again. It began at the same place and went on from chorus to verse. When she hummed an accompaniment to the music, the tempo was what one would have expected."[12]

The patients were always amazed to re-

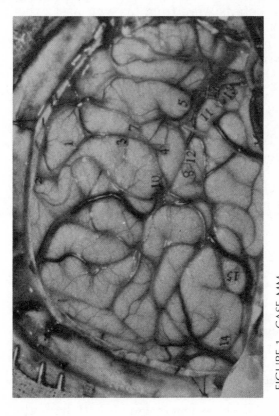

FIGURE 1. CASE MM
Right hemisphere exposed. The numbered tickets mark points where there were responses to the surgeon's stimulating electrode.

member the past in such vivid detail and assumed at once the surgeon was responsible for provoking the otherwise inaccessible memory. Each patient recognized that the details were from his own past experiences and it was clear that only those things he had paid attention to were preserved in the archives of the brain.

Occasionally, Penfield would warn a patient he was going to stimulate and then not do so. In such cases the patient reported no reaction at all.

Touching the speech area of the brain produced temporary loss of speech (aphasia) in the patient. Since the brain is not sensitive, the patient is unaware that he is aphasic until he tries to speak, or to understand speech, and is unable to do so.[13] Penfield relates what took place on one occasion:

"One of my associates began to show the patient a series of pictures on the other side of the sterile screen. C.H. named each picture accurately at first. Then before the picture of a butterfly was shown to him, I applied the electrode where I supposed the speech cortex to be. He remained silent for a time. Then he snapped his fingers as though in exasperation. I withdrew the electrode and he spoke at once: 'Now I can talk,' he said. 'Butterfly. I couldn't get that word butterfly, so I tried to get the word 'moth!'"[14]

The man understood with his mind what the card portrayed, and his mind called upon the speech center in the brain for the word corresponding to the concept present in his mind. The speech mechanism, therefore, is not identical with the mind, though it is directed by the mind. Words are expressions of thoughts, not the thoughts themselves. When the patient drew a blank because of the blockage of his speech center, he was puzzled and ordered the search for the name of a similar object, "moth." When that too failed, he snapped his finger in exasperation (this motor action not being controlled by the speech center). Then, finally, when his speech center was unblocked, he explained the entire experience using words appropriate to his thoughts. *"He* got the words from the speech mechanism when *he* presented concepts to it," Penfield concludes. "For the word 'he' in this introspection, one may substitute the word *mind*. Its action is not automatic."[15]

Penfield found similar results in brain areas controlling movements:

"When I have caused a patient to move his hand by applying an electrode to the motor cortex of one hemisphere, I have often asked him about it. Invariably his response was: 'I didn't do that. You did.' When I caused him to vocalize, he said: 'I didn't make that sound. You pulled it out of me.'"[16]

These involuntary movements are like a patient's leg jumping in response to the tap of a physician's hammer. Everyone recognizes that such movements are not acts of the will. Penfield summarizes:

"The electrode can present to the patient various crude sensations. It can cause him to turn head and eyes, or to move the limbs, or to vocalize and swallow. It may recall vivid re-experience of the past, or present to him an illusion that present experience is familiar, or that the things he sees are growing large and coming near. But he remains aloof. He passes judgment on it all. He says 'things are growing larger' but he does not move for fear of being run over. If the electrode moves his right hand, he does not say 'I wanted to move it.' He may, however, reach over with the left hand and oppose his action."[17]

As a result of observing hundreds of patients in this way, Penfield concludes, "The patient's mind, which is considering the situation in such an aloof and critical manner, can only be something quite apart from neuronal reflex action . . . although the content of consciousness depends in large measure on neuronal activity, awareness itself does not."[18]

Using these techniques of observation, Penfield was able to construct a complete map of the brain areas responsible for speech, movement, and all the internal and external senses. But neither the mind nor the will could be located in any part of the brain. The brain is the seat of sensation, of memory, of the emotions, and of the power of movement, but apparently the brain is not the organ of either the intellect or the will. (See Figure 2.)

Penfield declares, "None of the actions that we attribute to the mind has been initiated by electrode stimulation or epileptic discharge." He adds, "There is no place in the cerebral cortex where electric stimulation will cause a patient to believe or to decide."[19] The electrode can evoke sensations and memories, but it cannot make the patient syllogize or do algebra. It cannot even produce in the mind the simplest elements of reasoning. The electrode can make the patient's body move, but it cannot make him want to move it. It cannot coerce the will. Evidently, then, the human intellect and the human will have no bodily organs.

The New Story, as a consequence, sees

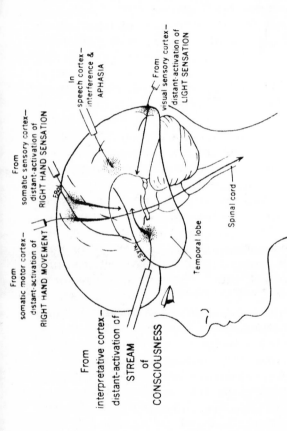

From
somatic motor cortex—
distant-activation of
RIGHT HAND MOVEMENT

From
somatic sensory cortex—
distant-activation of
RIGHT HAND SENSATION

In
speech cortex—
interference &
APHASIA

From
visual sensory cortex—
distant-activation of
LIGHT SENSATION

From
interpretative cortex—
distant-activation of
STREAM
of
CONSCIOUSNESS

Fiss.

VISS

Temporal lobe

Spinal cord

FIGURE 2. Activation of the Brain's Record of Consciousness and some other results of stimulation.

no impossibility in the will influencing matter. As Eccles explains,

"I have the indubitable experience that by thinking and willing I can control my actions if I so wish. . . . I am not able to give a scientific account of how thought can lead to action, but this failure serves to emphasize the fact that our present physics and physiology are too primitive for this most challenging task. . . . When thought leads to action, I am constrained, as a neuroscientist, to postulate that in some way, completely beyond my understanding, my thinking changes the operative patterns of neuronal activities in my brain. Thinking thus comes to control the discharges of impulses from pyramidal cells of my motor cortex and so eventually the contractions of my muscles and the behavioural patterns stemming therefrom."[20]

If the human will is non-material, then it is not unreasonable for it to act in ways matter cannot act; namely, by free choice. Consequently, the New Story sees nothing unscientific about the recognition in ourselves of the will's autonomy. Eccles concludes "there are thus no sound scientific grounds for denying the freedom of the will, which ironically, must be assumed if we are to act as scientific investigators."[21] In fact, a denial of free will would render the whole of science absurd. The scientist would have to ask not "What is true?" but "What are we conditioned to believe?" As physicist Carl von Weizsacker writes, "Freedom is a pre-requisite of the experiment. Only where my action and thought are not determined by circumstances, urges or customs but by my free choice can I make experiments."[22]

Moreover, the New Story sees nothing impossible in the mind's capacity to direct brain activities. Neuroscientist Roger Sperry describes the mentalist revolution that took place in psychology during the 1970's introducing a dramatic turnabout in the treatment of consciousness:

"Behaviorist principles, which had dominated for over half a century, were overturned. Psychology suddenly began to treat subjective events—mental images, inner thoughts, sensations, feelings, ideas, and so on—as factors having a genuine causal role in brain function and behavior. The contents of introspection, the whole world of inner experience, suddenly became accepted as elements that could influence physical and chemical events in the brain, they were no longer treated as passive, noncausal aspects or even as nonexistent ones."[23]

Sperry concludes:

"The higher cerebral properties of mind and consciousness are in command. They envelop, carry, and overwhelm the physico-chemical details. They call the plays, exerting downward control over the march of nerve-impulse traffic. Our new model, mentalism, puts the mind and mental properties to work and gives them a reason for being and for having evolved in a physical system."[24]

Knowledge and command require a certain distance. The mind cannot be an epiphenomenon of nerve machinery if it is to survey and direct the whole. For as Penfield says, "It is the mind (not the brain) that watches and at the same time directs."[25] The mind is responsible for the unity we experience in all our actions, thoughts, sensations, and emotions. Eccles adds, "The unity of conscious experience is provided by the self-conscious mind and not by the neuronal machinery."[26]

Again, if the brain is a supremely complex computer, then, like a computer, it must be directed by a mind. Penfield argues, "A computer (which the brain is) must be programmed and operated by an agency capable of independent understanding." Penfield spells out the role of the mind: "It is what we have learned to call the mind that seems to focus attention. The mind is aware of what is going on. The mind reasons and makes new decisions. It understands. It acts as though endowed with an energy of its own. It can make decisions and put them into effect by calling upon various brain mechanisms."[27] Thus, expecting the mind to be found in some part of the brain, or in the whole brain, is like expecting the programmer to be part of the computer.

On the strength of the foregoing evidence, Penfield sees no hope for the Old Story's materialistic approach to mind: "To expect the highest brain mechanism or any set of reflexes, however complicated, to carry out what the mind does, and thus perform all the functions of the mind, is quite absurd."[28] Biologist Adolf Portmann agrees that "no amount of research along physical . . . or chemical lines can ever give us a full picture of psychological, spiritual or intellectual processes."[29]

Nor does Penfield foresee the eventual derivation of mind from matter by some future physiology as was the expectation in the Old Story: "It seems to be certain that it will always be quite impossible to explain the mind on the basis of neuronal action within the brain." This is why he maintains it is "more reasonable to suggest . . . that the mind may be a distinct and *different* essence" from the body.[30]

Ironically, Penfield began his investigations with the intention of proving just the opposite: "Through my own scientific career," he remarks, "I, like other scientists, have struggled to prove that the brain accounts for the mind." He started with all the assumptions of the Old Story. But the evidence finally compelled Penfield to admit that the human mind and the human will are non-material realities. "What a thrill it is, then," he declares, "to discover that the scientist, too, can legitimately believe in the existence of the spirit!"[31] And if the mind and the will are non-material, then these faculties are, in the words of Eccles, arguably "not subject in death to the disintegration that affects . . . both the body and the brain."[32]

III

BEAUTY

Physicist Louis de Broglie writes, "In every epoch in the history of science, aesthetic feeling has been a guide that has directed scientists in their research." Beauty has always been a central principle in science. According to the Old Story, however, matter has only quantitative properties such as weight, size, shape, and number. Since beauty is not among these, the Old Story tends to consider it a property of the observer rather than a quality of natural things. "Neither the beautiful nor the pleasant signifies anything other than the attitude of our judgment to the object in question," writes Descartes in 1630. Spinoza agrees: "Beauty . . . is less a quality of the object studied than the effect arising in the man studying that object."[1]

These and other thinkers initiated a strong current that persisted. Two centuries later Charles Darwin articulates the Old Story's view of beauty when he writes, "The sense of beauty obviously depends on the nature of the mind, irrespective of any real quality in the admired object." Freud felt compelled to reduce beauty to instinct: "Psychoanalysis, unfortunately, has scarcely anything to say about beauty . . . All that seems certain is its derivation from the field of sexual feeling."[2]

If beauty is not a quality of nature, as the Old Story holds, then two things follow. First, beauty, though perhaps privately enjoyable, cannot enter into a scientific argument. Beauty will not be of any particular help in discovering the truths of nature. Second, the fine arts, to the extent that they pursue beauty, cannot have anything in common with the sciences. In the Old Story, the sciences are sometimes caricatured as cold-hearted but hard-headed and the arts as warm-hearted but soft-headed. Entomology is expected to be as silent about the beauty of butterflies as poetry is about their digestive enzymes.

In the New Story, by contrast, beauty is a means of discovering scientific truth. For instance, James Watson in his book *The Double Helix,* mentions how beauty guided the discovery of DNA's molecular make-up: "So we had lunch, telling each other that a structure this pretty just had to exist." And "almost everyone . . . accepted the fact that the structure was too pretty not to be there."[3]

All of the most eminent physicists of the twentieth century agree that beauty is the primary standard for scientific truth. According to physicist Richard Feynman, in science "you can recognize truth by its beauty and simplicity." And Heisenberg declares that beauty "in exact science, no less than in the arts . . . is the most important source of illumination and clarity."[4]

By looking to beauty the great theoretical physicists of our age have made major discoveries. Concerning quantum mechanics in which he pioneered, Werner Heisenberg remarks that it was "immediately found convincing by virtue of its completeness and abstract beauty." General relativity is considered by physicists as probably the most beautiful of all existing physical theories. Erwin Schrödinger gives it this tribute: "Einstein's marvellous theory of gravitation . . . could only be discovered by a genius with a strong feeling for the simplicity and beauty of ideas." And Einstein himself referred to its beauty at the end of his first paper on gravitation: "Scarcely anyone who fully understands this theory can escape its magic."[5]

Beauty is so central a standard in physics that it takes primacy even over experiment. Physicist Paul Dirac states "It is more important to have beauty in one's equations than to have them fit experiment." This can be understood if we picture the theoretician before a bewildering mass of experimental data. Which results are the important ones? How should they all be interpreted? What is the pattern? Here beauty is a trustworthy guide. Physicist Sir George Thomson writes: "One can always make a theory, many theories, to account for known facts, occasionally even to predict new ones. The test is aesthetic. Some theories are cumbrous, limited in scope and arbitrary. They seldom live long."[6]

Beauty even challenges "the facts". A striking illustration is found in a scientific paper presented by physicists Richard Feynman and Murray Gell-Mann, in 1958, proposing a new theory of weak interactions. The theory boldly contradicted a number of experiments. Its main attraction was its beauty. Feynman and Gell-Mann argued "It's universal, it's symmetric . . . it is the simplest possibility," and this "indicates that these experiments are wrong." Gell-Mann comments:

"Frequently a theorist will *throw out* a lot of data on the grounds that if they don't fit an elegant scheme, they're wrong. That's happened to me many times. The theory of weak interactions: there were nine experiments that contradicted it—all wrong. Every one. When you have something simple that agrees with all the rest of physics and really seems to explain what's going on, a few experimental data against it are no objection whatever. Almost certain to be wrong."[7]

In physics beauty reigns supreme. Experiment often errs, beauty seldom. If occasionally a supremely elegant theory does not fit one group of facts, it inevitably finds application elsewhere. For instance, during the twenties, mathematician-physicist Hermann Weyl became convinced his gauge theory was inapplicable to gravity. But, because of its aesthetic perfection, he did not wish to abandon it entirely. Much later, Weyl's theory was found to illuminate quantum electrodynamics, confirming his aesthetic judgment.[8]

Far from being "unscientific", beauty animates science. And the beauty physicists seek is not the product of private or idiosyncratic emotion. On the contrary, the physicists themselves indicate three specific elements of beauty. Einstein summarizes these three elements of scientific beauty in one formula: "A theory is the more impressive the greater the simplicity of its premises is, the more different kinds of things it relates, and the more extended is its area of applicability."[9] Simplicity, then, is the first element of beauty. The "different kinds of things it relates" means how the theory harmonizes disparate things. Thus, we may label the second element harmony. And the extended applicability is a theory's brilliance; that is, how much clarity it has in itself and how much light it sheds on other things. Gell-Mann's formula, cited above, captures the three aspects of beauty in a single concise phrase: "something simple that agrees with all the rest of physics and really seems to explain what is going on." Simplicity, harmony, and brilliance. Each of these calls for a brief explanation.

Simplicity. There exist today other theories of gravitation besides Einstein's but because they lack simplicity, none of them are taken seriously. "Most rival theories are convincingly disproved," observes astrophysicist Roger Penrose, "the few that remain having been, for the most part, contrived directly so as to fit with those tests that have been actually performed. No rival theory comes close to general relativity in elegance or simplicity of assumption."[10]

The principle of simplicity implies two things—completeness and economy. Mathematician-physicist Henri Poincaré tells us "It is because simplicity and vastness are both beautiful that we seek by preference simple facts and vast facts." A theory beautiful by this standard must take into account all the facts and must include only what is necessary. Nothing lacking, nothing superfluous. A demanding standard indeed! Heisenberg says of quantum theory that it was "immediately found convincing by virtue of its completeness and abstract beauty."[11]

Harmony. Einstein declares "without the belief in the inner harmony of the world there could be no science." Heisenberg describes harmony as the "proper conformity of the parts to one another and to the whole." In any science a good theory will harmonize many previously unrelated facts. Harmony also implies symmetry. There is a pleasing symmetry to all the laws of physics. "Every law of physics . . . goes back to some symmetry of nature," writes Wheeler. Heisenberg adds: "The symmetry properties always constitute the most essential features of a theory." Newton's third law is a well-known example of symmetry in physics: "To every action there is always opposed an equal reaction." This mirror symmetry is found on the subatomic level, where to every kind of particle there corresponds an anti-particle with the same mass but with opposite characteristics. In fact, the successful prediction of the existence of many subatomic particles was made primarily on the basis of this symmetry.[12]

Brilliance. A theory with this quality has great clarity in itself and sheds light on many other things, suggesting new experiments. Newton, for example, astounded the world by explaining falling bodies, the tides, and the motions of the planets and the comets with three simple laws. George Thomson states, "in physics, as in mathematics, it is a great beauty if a theory can bring together apparently very different phenomena and show that they are closely connected; or even different aspects of the same thing." General relativity does precisely that in an elegant and surprising way, as astrophysicist S. Chandrasekhar points out: "It consists primarily in relating, in juxtaposition, two fundamental concepts which had, till then, been considered as entirely independent: the concepts of space and time, on the one hand, and the concepts of matter and motion on the other." Moreover, general relativity has proven extraordinarily brilliant, shedding its light on cosmology, astrophysics, and quantum mechanics.[13]

The Old Story maintains that simplicity and the other elements of beauty are not laws of nature but at most laws of the human mind. Newton does not agree, but ascribes simplicity to nature, not to man: "Nature is pleased with simplicity, and af-

fects not the pomp of superfluous causes."
And the testimony of twentieth-century
physicists is clear in this matter. Feynman
declares, "Nature has simplicity and therefore a great beauty." He does not ascribe
the beauty to the onlooker. And Wheeler
asserts, "Every law of physics ... goes
back to some symmetry of nature," not
back to a symmetry of our minds. And Max
Born affirms, "The genuine physicist believes obstinately in the simplicity and
unity of nature, despite any appearance to
the contrary." In a conversation with Einstein, Heisenberg once said:

"I believe, just like you, that the simplicity of natural laws has an objective character, that it is not just the result of thought
economy. If nature leads us to mathematical forms of great simplicity and beauty
... we cannot help thinking they are true,
that they reveal a genuine feature of
nature."[14]

Chandrasekhar adds, "All of us are sensitive to nature's beauty. It is not unreasonable that some aspects of this beauty
are shared by the natural sciences." Again,
the source of the beauty is nature, not
man. Why is beauty found in natural science? Because nature is filled with beauty.
Physicist David Bohm declares, "Almost
anything to be found in nature exhibits

some kind of beauty both in immediate perception and in intellectual analysis." Henri Poincaré says, "The scientist does not study nature because it is useful to do so. He studies it because he takes pleasure in it; and he takes pleasure in it because it is beautiful. If nature were not beautiful, it would not be worth knowing and life would not be worth living."[15] And Carl von Weizsacker adds an explanation, arguing that "the often-cited principle of economy of thought explains, at the most, why we look for simple laws, but not why we find them."[16] The New Story, then, proposes beauty as a standard in science because nature is beautiful. On this view, a beauty-blind scientist would be a poor scientist.

Because the physics of the New Story acknowledges beauty as a property of nature, it opens up a common ground between the sciences and the fine arts. Physicist and novelist C.P. Snow, after experience in both science and art, is unusually qualified to speak of the beauty found in both domains:

"Anyone who has ever worked in any science knows how much aesthetic joy he has obtained. That is, in the actual *activity* of science, in the process of making a discovery, however humble it is, one can't help feeling an awareness of beauty. The

subjective experience, the aesthetic satisfaction, seems exactly the same as the
satisfaction one gets from writing a poem
or a novel, or composing a piece of music. I
don't think anyone has succeeded in distinguishing between them. The literature
of scientific discovery is full of this aesthetic joy. The very best communication of
it that I know comes in G.H. Hardy's book,
A Mathematician's Apology. Graham
Greene once said he thought that, along
with Henry James's prefaces, this was the
best account of the artistic experience ever
written."[17]

The elements of beauty are found not
only in physics but also have their
analogues in the arts. Simplicity, for
example, is clearly a goal of the artist.
Great works of art are generally recognized as fulfilling the exacting standard of
nothing lacking, nothing superfluous.
Albrecht Dürer certainly has the principle
of simplicity in mind when he advises artists: "There is a right mean between too
much and too little; strive to hit upon this
in all your works." And Vincent Van Gogh
praises the simplicity and economy of the
Japanese artists: "Their work is as simple
as breathing. And they do a figure in a few
strokes with the same ease as if it were as
simple as buttoning your coat. Oh! I must
manage some day to do a figure in a few

strokes." Johannes Brahms speaks of the difficulty of achieving simplicity in music: "It is not hard to compose, but it is wonderfully hard to let the superfluous notes fall under the table." Just as a beautiful scientific theory is vast yet simple, so a fine painting expresses a vast range of experience in a simple way. As Henri Matisse puts it, "I want to reach that state of condensation of sensations which constitutes a picture."[18]

Along with the second element, harmony, we can include the symmetry and proportion mentioned by the physicists. Dürer remarks, "Without just proportion, no figure can be perfect, no matter how diligently it might be executed." To verify symmetry and balance, the Renaissance sculptor and artist Alberti recommended looking at a painting's reflection: "It is wonderful how any defect in a painting shows its ugliness in the looking glass. Therefore things drawn from nature are to be amended with a mirror." A child's kaleidoscope also demonstrates how reflected symmetry can render beautiful what would otherwise be dull.[19]

Eighteenth-century composer Christoph Gluck compared harmony in music to proportion in a drawn figure: "The slightest alteration in outline, that could in no way destroy the likeness in a caricature, can entirely disfigure the portrait of a lovely woman . . . and the greatest beauties of melody and harmony become defects and imperfections when used out of place." Painters speak of the friendship of certain colors and their natural harmonies. Matisse describes his goal in painting: "When I have found the relationship of all the tones, the result must be a living harmony of tones, a harmony not unlike that of a musical composition . . . Until I achieve this proportion in all the parts of the composition, I strive towards it and keep on working." The aim of Matisse is analogous to the aim of a theoretical physicist who strives to harmonize all the data in the simplest way.[20]

The last element of beauty is brilliance. "Light," says Edouard Manet, "is the principal personage of a painting." Leonardo da Vinci, in his handbook on painting, suggests sketching persons seated in the doorway of a dark house: "This manner of treating and intensifying light and shadow adds much to the beauty of faces."[21] Light, however, takes on a special splendor and rich brilliance when it is divided into color—thus the beauty of sunsets, rainbows, tropical fish, butterflies, and flowers. One reason the impressionist paintings of the late nineteenth century are admired is that they emphasize the special beauty of light and color.

As regards music, the clarity of a sound

is an undeniable element of its beauty. And "timbre in music," writes Aaron Copland, "is analogous to color in painting." Timbre or tone color enables the ear to distinguish a flute from a trumpet, even when both instruments play the same note. In the nineteenth century, composers began to use tone color to produce a musical brilliance similar to the brilliance of visual color found in impressionist paintings. Rimsky-Korsakov comments on this movement in music: "Our post-Wagnerian epoch is the age of brilliance and imaginative quality in orchestral tone-coloring. Berlioz, Glinka, Liszt, Wagner . . . and others have brought this side of musical art to its zenith; they have eclipsed, as colorists, their predecessors." Both brilliance and color, then, are elements in the beauty of painting and music.[22]

Physicist Steven Weinberg stresses the new unity between science and the fine arts:

"Science has certain parallels with art. For one thing, there is the quest by scientists for beauty and simplicity. We pursue this because we believe that the fundamental laws underlying nature have to be simple; so we look for simplicities that might mirror the ultimate simplicities of the laws of nature. In my field— elementary-particle physics—we have discovered that nature is far more simple than it appears superficially. The simplicity takes the form of principles of symmetry. For example, there's a deep symmetry between two elementary particles—the neutrino and the electron—that superficially appear to have quite different properties."[23]

Weinberg points to a further parallel in the way scientists and artists work: "Scientists, like artists, rely heavily on intuition. Very often I turn my back on a whole line of work because it just doesn't feel right to me, or I'll spend months developing a line of work because it has just the right feeling."[24]

Thus, the simplicity, harmony, symmetry, proportion, brilliance, and clarity seen in the most beautiful physical theories have their analogous counterparts in the beauty found in painting and music. It is not difficult to see that the same standards of beauty would also apply to poetry, the dance, and the other arts. The New Story indicates that the elements of the invisible, intellectual beauty of physics are analogous to the elements that characterize visible and audible beauty in the fine arts. In the New Story, the scientist and the artist pursue the same goal of beauty by different paths. "It is probably correct," writes Heisenberg, "to say that the world of poetry has been familiar to all really great scientists. The physicist, at any rate, also needs to discover the harmonies of natural events."[25]

IV

GOD

According to the Old Story of science, matter is central and mind is secondary. This axiom is illustrated by the way the Old Story looks at man in the world: the human mind is dwarfed into insignificance by an infinite, impersonal universe of matter. Compared to the earth, or to the sun, or to the galaxy, man is negligible in size and inconsequential in power. The Old Story criticizes what it considers to be man's exaggerated sense of self-importance, arguing that Copernicus has dethroned conceited man from the center of the universe. Man should realize that he is a tiny inhabitant on an insignificant planet circling an ordinary star.

The logic of materialism rejects purpose. If the universe is nothing but matter, there can be no purpose in natural things. For matter cannot intend anything; it cannot plan; it acts only by internal, mechanical necessity. Thus scientific explanations must employ only material and mechanical causes. Both Bacon and Descartes exclude from natural science any appeal to purpose. Bacon writes: "The final cause [purpose] rather corrupts than advances the sciences." And Descartes says the same: "Every kind of cause that is customarily drawn from the end is of no use in things physical or natural." This rejection is for methodological reasons, for Bacon and Descartes were not materialists or atheists. But the consequence remains; there is no role for purpose in Old Story science.[1]

Is there room for God in such a universe? Newton thought so and tried to reserve a place for a deity in his mechanical system of the heavens. In a letter to Dr. Richard Bentley in 1692, Newton maintained that God was necessary to establish the motion of the planets and the original structure of the solar system: "The motions, which the planets now have, could not spring from any natural cause alone, but were impressed by an intelligent Agent."[2]

If Newton was unwilling to leave God out of science, others were not. Thus the famous reply of mathematician-astronomer Pierre-Simon Laplace when Napolean asked him about the place of God in his celestial mechanics: "Sire, I have no need of that hypothesis." By the nineteenth century many considered God to be invisible to the mind as well as to the eye. Science became progressively more agnostic. In fact, the word *agnostic* was coined in 1869 by biologist Thomas Huxley. Carl F. Gauss, the greatest mathematician of the nineteenth century, reflects the doubt: "There are problems to whose solution I would attach infinitely greater importance than to those of mathematics, for example touching ethics, or our relation to God, or concerning our destiny and our future; but their solution lies wholly beyond us and completely outside the province of science."[3]

Other thinkers held that science does not allow agnosticism. They argued that since the universe is a self-running machine, it needs no supernatural cause at all. If matter is eternal, there seems to be no need for a Creator. Atheism was thus considered by many to be more honest and more consistent with the Old Story of science.

Freud is one representative of an Old Story attitude toward religion. "The religions of mankind," he states, "must be classed among the mass delusions . . . ." In religion man seeks an escape from reality. "Religious ideas," Freud continues, "have arisen from . . . the necessity of defending oneself against the crushingly superior force of nature." According to Freud, men tend to believe in a father behind the universe because, as infants, they had so much need of a father's care. Thus man creates God, not the reverse. Freud adds that men "will have to admit to themselves the full extent of their helplessness and their insignificance in the machinery of the universe; they can no longer be the center of creation, no longer the object of a beneficent Providence." And, concerning religion, Freud predicts "surely this infantalism is destined to be surmounted." Man must have courage enough to recognize that he is alone in a vast and impersonal universe.[4]

The Old Story produced in the nineteenth century some especially ardent works attacking religion in the name of science. Just to give a sample, 1875 saw the publication of *The History of the Conflict Between Religion and Science* by John W. Draper, the first president of the American Chemical Society. Twenty years later, in 1895, Andrew D. White, the first president of Cornell University, wrote *A History of the Warfare of Science with Theology in Christendom*. The titles alone are sufficient to indicate the trend.

What does the New Story have to say about all this? First of all, the New Story of science has exciting news about the universe itself: "Perhaps the most important scientific discovery of the twentieth century," writes astrophysicist Dennis Sciama, "is that the Universe as a whole, considered as a single totality, is amenable to rational inquiry by the methods of physics and astronomy." This new perspective on the universe became possible with the advent of Einstein's theory of general relativity, which, unlike Newtonian physics, integrated gravity, space, and time. "Space, . . . Einstein taught us, is a participant in physics and not only an arena for physics," writes Wheeler.[5] The same holds for time. This integration for the first time gave physicists the tools to investigate in detail the structure, the origin, and the destiny of the whole universe. After publication of general relativity, astronomer Willem de Sitter and mathematician Alexander Friedmann independently deduced from the new theory that the universe is expanding. This was soon confirmed by observation. During the 1920's, astronomer Edwin Hubble, analyzing the light from distant galaxies, discovered that all the observable galaxies are receding from each other. This was the first clue to the history of the universe. If the galaxies are receding from each other now, then a very long time ago they must have been together. This argues a beginning.

A second clue came from nuclear

physics. Nineteenth century chemists knew that the sun could not be burning conventional fuel. Ordinary chemical combustion could not account for the energy of the sun, for if the whole mass of the sun were coal, it would burn itself out in 300 years. The sun remained a mystery until the discovery of nuclear energy in the early years of the twentieth century. Finally, in 1938, physicists Hans Bethe and Carl von Weizsacker put together a full account of how the sun produces energy by a nuclear transmutation of the elements. In the core of the sun hydrogen is converted into helium, producing energy and light. Over millions of years, nuclear processes within each star slowly build up not only helium but also all the heavier elements: carbon, oxygen, silicon, iron, and the rest. Now if all the heavy elements in the universe have been manufactured out of hydrogen in the cores of stars, then the universe must have been originally composed almost entirely of hydrogen. This again argues a beginning.

Finally, in 1948, physicist George Gamow, synthesizing the evidence from the receding galaxies and from the life cycle of stars, proposed that the universe itself resulted from a primordial expansion of matter which he dubbed "The Big Bang". Like an explosion, the intensely hot fireball would have rapidly expanded and cooled down. Using nuclear physics, Gamow showed how the subatomic particles present at the earliest stages produced, at later temperatures and pressures, the atoms of the newborn universe. Furthermore, he showed that as a consequence of the expanding and cooling processes, a faint afterglow of radiation ought to be dispersed uniformly throughout the universe.

Gamow's prediction lay dormant for several years. But then, in 1965, quite by accident, Arno Penzias and Robert Wilson, using a huge microwave receiver, discovered a faint radiation emanating from space. Measuring this radiation with unprecedented accuracy, Penzias and Wilson found it to be about 3.5° above absolute zero. It was not more intense in the direction of the sun or in the direction of the Milky Way. It could not, therefore, be of solar or galactic origin. Only one explanation remained: it was the relic background radiation of the Big Bang. This observational evidence confirmed the Big Bang.

Our universe, then, is the aftermath of a gigantic expansion of matter. Present size and rate of expansion indicate that the universe began about 12 to 20 billion years ago. At one sextillionth of a sextillionth of a second after the beginning, all the matter in the universe was packed into a space much, much smaller than a proton occupies. The density at that point staggers the imagination. Imagine—planets, stars, whole galaxies, all the matter and energy in the universe—contained in a space next to nothing in size! And at time zero the density was infinite with no extension in space at all. That instant marked the beginning of matter, time, and space. The Big Bang must not be pictured as the expansion of matter within an already existing space. The Big Bang is the expansion of space itself. This can be understood by the intellect but not pictured by the imagination.

Advocates of the Old Story were uncomfortable with the idea of an absolute beginning. So they devised alternative cosmologies to salvage the eternity of matter. One of the alternatives, the steady-state hypothesis, was formulated by astronomer Sir Fred Hoyle and it required the spontaneous generation of hydrogen throughout the universe. The discovery of the universe's background radiation, however, conclusively eliminated this alternative.

Another model proposed to avoid a beginning was the oscillating universe. If there is sufficient matter in the universe, the pull of gravity will eventually halt the present expansion and reverse it, resulting in the eventual re-collapse of all matter in what might be called "The Big Crunch". The oscillating theory suggests that another Big Bang might follow the collapse and that the universe might thus oscillate forever between Bangs and Crunches. In this way the eternity of matter could be retained. Steven Weinberg, author of *The First Three Minutes,* a definitive account of the early stages of the universe, comments on this theory:

"Some cosmologists are philosophically attracted to the oscillating model, especially because, like the steady-state model, it nicely avoids the problem of Genesis. It does, however, face one severe theoretical difficulty. In each cycle the ratio of photons to nuclear particles (or, more precisely, the entropy per nuclear particle) is slightly increased by a kind of friction (known as "bulk viscosity") as the universe expands and contracts. As far as we know, the universe would then start each new cycle with a new, slightly larger ratio of photons to nuclear particles. Right now this ratio is large, but not infinite, so it is hard to see how the universe could have previously experienced an infinite number of cycles."[6]

Weinberg's argument here is based on an inevitable consequence of one of the most fundamental properties of matter: the second law of thermodynamics. This law says that if matter is compressed it will heat up and its entropy will increase. Thus the more Big Crunches the universe has gone through, the greater its tempera-

ture and entropy will have to be. And since
the universe right now is not infinite in
temperature or in entropy, it must have
had a beginning. In an oscillating universe, each Big Bang would have begun at
a higher temperature than the preceding
one. Hence a universe at the end of a long
series of Big Bangs and Big Crunches
would have to be much hotter than 3.5°
absolute.

In fact, detailed thermodynamic arguments indicate no re-expansion at all.
Physicist Sidney A. Bludman: "Our Universe cannot bounce in the future. Closed
Friedman universes were once called oscil-
latory universes. We now appreciate that,
because of the huge entropy generated in
our Universe, far from oscillating, a closed
universe can only go through one cycle of
expansion and contraction. Whether closed
or open, reversing or monotonically expanding, the severely irreversible phase
transitions give the Universe a definite
beginning, middle and end." Nor does the
oscillating model harmonize with general
relativity. Hence, John Wheeler concludes
that a single Big Crunch would end the
universe forever: "With gravitational collapse we come to the end of time. Never out
of the equations of general relativity has
one been able to find the slightest argument for a 're-expansion' or a 'cyclic universe' or anything other than an end."[7]

It appears then that matter is not eternal after all. As astrophysicist Joseph Silk declares, "The beginning of time is unavoidable." And astronomer Robert Jastrow concludes, "The chain of events leading to man commenced suddenly and sharply at a definite moment in time, in a flash of light and energy."[8]

Is there room for a God in such a universe? Physicist Edmund Whittaker thinks so. He argues "There is no ground for supposing that matter and energy existed before [the Big Bang] and were suddenly galvanized into action. For what could distinguish that moment from all other moments in eternity? It is simpler to postulate creation *ex nihilo*—Divine will constituting Nature from nothingness." And physicist Edward Milne, reflecting on the expanding universe, concludes, "As to the first cause of the Universe, in the context of expansion, that is left for the reader to insert, but our picture is incomplete without Him."[9]

In the New Story of science the whole universe—including matter, energy, space, and time—is a one-time event and had a definite beginning. But something must have always existed; for if ever absolutely nothing existed, then nothing would exist now, since nothing comes from nothing. The material universe cannot be the thing that always existed because matter had a beginning. It is 12 to 20 billion years old. This means that whatever has always existed is non-material. The only non-material reality seems to be mind (see Chapter II). If mind is what has always existed, then matter must have been brought into existence by a mind that always was. This points to an intelligent, eternal being who created all things. Such a being is what we mean by the term God.

But what response does the New Story have to the Old Story's claim that man is dwarfed into insignificance by the vastness of the universe, that he can "no longer be the center of creation, no longer the object of a beneficent Providence"?[10]

The New Story begins by correcting a misunderstanding. Astrophysicist Brandon Carter of Cambridge writes:

"Copernicus taught us the very sound lesson that we must not assume gratuitously that we occupy a privileged *central* position in the Universe. Unfortunately there has been a strong (not always subconscious) tendency to extend this to a most questionable dogma to the effect that our situation cannot be privileged in any sense."[11]

Carter argues that certain initial conditions of the universe surprisingly favorable to life should be considered "as confirming 'conventional' (General Relativistic, Big Bang) physics and cosmology which could in principle have been used to predict them all in advance of their observation. However these predictions do require the use of what may be termed the *anthropic principle* to the effect that what we can expect to observe must be restricted by the conditions necessary for our presence as observers. (Although our situation is not necessarily *central*, it is privileged to some extent.)"[12]

Astrophysicist Steven Hawking of Cambridge refers to the Anthropic Principle in addressing the question why the universe should be expanding at just the correct rate to avoid recollapse:

"The only 'explanation' we can offer is one based on a suggestion of Dicke (1961) and Carter (1970). The idea is that there are certain conditions which are necessary for the development of intelligent life: out of all conceivable universes, only in those in which these conditions occur will there be beings to observe the Universe. Thus our existence requires the Universe to have certain properties. Among these properties would seem to be the existence of gravitationally bound systems such as stars and galaxies and a long enough time-scale for biological evolution to occur. If the Universe were expanding too slowly, it would not have this second property for it would recollapse too soon. If it were expanding too fast, regions which had slightly higher densities than the average or slightly lower rates of expansion would still continue expanding indefinitely and would not form bound systems. Thus it would seem that life is possible only because the Universe is expanding at just the rate required to avoid recollapse.

"The conclusion is, therefore, that the isotropy of the Universe and our existence are both results of the fact that the Universe is expanding at just about the critical rate. Since we could not observe the Universe to be different if we were not here, one can say, in a sense, that the isotropy of the Universe is a consequence of our existence."[13]

Hawking's argument can be read in two different ways. The Old Story would hold that anything in the universe conducive to life is coincidence and that "the universe, meaningless or not, would still come into being and run its course even if the constants and initial conditions forever ruled out the development of life and consciousness. Life is accidental and incidental to the machinery of the universe," as Wheeler characterizes the Old Story.[14]

The alternative is to look at the universe as aiming at life and at man. This corresponds to what Carter calls "the *strong anthropic principle* stating that the Universe (and hence the fundamental parameters on which it depends) must be such as to admit the creation of observers within it at some stage."[15] In this vein Wheeler asks:

"What possible sense it could make to speak of 'the universe' unless there was someone around to be aware of it. But awareness demands life. Life in turn, however anyone has imagined it, demands heavy elements. To produce heavy elements out of the primordial hydrogen requires thermonuclear combustion. Thermonuclear combustion in turn needs several times 10^9 years cooking time in the interior of a star. But for the universe to provide several times 10^9 years of time, according to general relativity, it must have a reach in space of the order of several times 10^9 . . . [light years]. Why then is the universe as big as it is? Because we are here!"[16]

A startling reversal of the Old Story perspective. The vastness of the universe is seen as making life possible.

Nor is the Anthropic Principle limited to cosmology. Physicist Freeman Dyson indi-

cates how even the forces that bind neut-
rons and protons in the nucleus of the atom
had to be what they are if life were to be
possible:

"If the nuclear forces had been slightly
stronger than they are, the diproton would
exist and almost all the hydrogen in the
universe would have combined into dipro-
tons and heavier nuclei. Hydrogen would
be a rare element, and stars like the sun,
which live a long time by the slow burning
of hydrogen in their cores, could not exist.
On the other hand, if the nuclear forces
had been substantially weaker than they
are, hydrogen could not burn at all and
there would be no heavy elements. [And
hence no life.] If, as seems likely, the evolution of life requires a star like the sun,
supplying energy at a constant rate for billions of years, then the strength of nuclear
forces had to lie within a rather narrow
range to make life possible."[17]

Many other instances might be cited.
Dyson notes, for example:

"If the laws were changed so that electrons no longer excluded each other, none
of our essential chemistry would survive.
There are many other lucky accidents in
atomic physics. Without such accidents,
water could not exist as a liquid, chains of

carbon atoms could not form complex organic molecules, and hydrogen atoms could not form breakable bridges between molecules."[18]

The properties of matter, then, on the smallest scale and on the scale of the whole universe appear uniquely suited to life. Not only are there many instances but in each case a slight increase or decrease in the parameter would render life impossible. Wheeler speaks of the ensemble of possible universes, but points out that only a tiny fraction of them could support life.[19] Dyson, surveying this broad pattern, concludes that it argues purpose, not coincidence: "The more I examine the universe and study the details of its architecture, the more evidence I find that the universe in some sense must have known we were coming."[20] Certain conditions necessary for life were built into the Big Bang from the very beginning.

Wheeler stresses that "no reason has ever offered itself why certain of the constants and initial conditions have the values they do except that otherwise anything like observership as we know it would be impossible." Consequently, rather than resolving the issue to coincidence, he asks whether there is not greater likelihood that "no universe at all could come into being unless it were guaranteed to produce life, consciousness and observership somewhere and for some little length of time in its history-to-be?" Life is not accidental. On the contrary, Wheeler asserts that "Quantum mechanics has led us to take seriously and explore the directly opposite view that the observer is as essential to the creation of the universe as the universe is to the creation of the observer." Though man is not at the physical center of the universe, he appears to be at the center of its purpose. According to Erwin Schrödinger, without man the universe would be a drama played before empty stalls.[21]

A universe aiming at the production of man implies a mind directing it. For matter on its own cannot aim at anything. Hence, the New Story again leads to a mind that directs the whole universe, all the laws of nature and all the properties of matter, to a goal. To that mind we give the name God. Heisenberg describes the Old Story's methodology: "The mechanics of Newton and all the other parts of classical physics constructed after its model started from the assumption that one can describe the world without speaking about God or ourselves."[22] A universe without mind. But the New Story indicates the opposite on both counts. The Big Bang and the Anthropic Principle point to minds at both ends of the universe.

The New Story not only affirms the primacy of mind in the universe, it also affirms that beauty is part of the way the world is put together (see Chapter III). This fresh perspective on beauty also leads to evidence for God.

Nature abounds with beauty. In the inanimate world, for instance, naturally occurring geodes, precious stones, and crystals exhibit an undeniable beauty of symmetry, color, and brilliance. A striking example of such beauty is the snowflake. Figure 3 illustrates the amazing variety of snowflake patterns, all based on the hexagon. The twelve snowflakes shown appear in *Snow Crystals,* a book of more than 2000 snowflakes painstakingly photographed by W.A. Bentley over a period of almost fifty years. W.J. Humphreys introduces the book with the following reflection:

"Snow, the beautiful snow, as the raptured poet sang, winter's spotless downy blanket for forest and field, has ever challenged pen to describe, and brush to paint, its marvelous mass effects. Nor is the aesthetic urge of its very tiniest flake or smallest crystal that gently floats from heaven to earth any less compelling. It is even more insistent—doubly more—for it not only quickens that response to the dainty and the exquisite that makes us human, but equally arouses our desire to under-

FIGURE 3. Snowflakes

stand, our curiosity to know, the how and the why of this purest gem of surpassing beauty and of a myriad of forms."[23]

Textile designers and artists derive inspiration from Bentley's catalogue of snowflakes, drawing upon what Humphreys calls the "endless gallery of Nature's delicate tracery and jewel design."[24]

Even with the smallest snow storm trillions of flakes fall to the earth. Each is probably unique. No one has yet worked out the complete physics of how snowflakes are formed, although James Langer of the Institute for Theoretical Physics at Santa Barbara has developed a promising mathematical model after many years of labor.[25]

Can the mechanisms of nature account for the beauty of snowflakes, or sea foam, or rainbows, or sunsets? The beauty of these inanimate things follows by necessity from the laws of physics and chemistry, which laws themselves are beautiful, as we saw in Chapter III. Given those laws of nature, an ugly universe could never ensue. The beauty of inanimate things is built right into the machinery of nature. To take an analogy, one might construct a completely mechanized automobile factory that produced beautiful vehicles. One could build the resulting beauty of design and color right into the machinery. But beauty in an automobile does not thereby become an absolute necessity. Ugly vehicles could still transport passengers efficiently. And machines could be invented to produce them. In the same way, no absolute necessity requires nature's physical laws to incorporate simplicity and symmetry in the first place. Some other universe with asymmetrical, needlessly complex physical laws could conceivably produce ugly snowflakes by mechanical necessity.

Necessity, then, yields no ultimate explanation of the beauty found in non-living things. Neither can it explain the beauty found in plants and animals. Biologist Adolf Portmann, recognized authority on the shapes and markings of living things, points to many features incomprehensible in terms of necessity. Leaves are necessary for a tree to produce its food, Portmann points out, "yet how much in the shape and outline of a leaf is not adaptation to the environment but pure self-representation!" The requirements of photosynthesis explain why a tree has leaves in the first place but not why a maple leaf is different from an oak leaf.[26]

The same holds for animals. For instance, concerning the feathers of birds, Portmann remarks, "For a long time, feathers were thought to play no other role than to facilitate heat regulation and flight. However, we must now introduce a third role: self-expression, for there are many feathers whose external structure is predominantly ornamental."[27]

The human body also demonstrates that necessity cannot account for beauty. The human voice is more versatile and expressive than any musical instrument. That man has a voice capable of producing beautiful sounds is not demanded by necessity; a dull monotone or a raucous screech would have sufficed to call for help or to communicate physical needs. Darwin himself recognized that necessity cannot explain man's musical endowments: "As neither the enjoyment nor the capacity of producing musical notes are faculties of the least use to man in reference to his daily habits of life, they must be ranked amongst the most mysterious with which he is endowed."[28] Necessity might explain why a bird call is beautiful to another bird, but not why it is beautiful to a man. By the same token, why should a leopard be beautiful to a man? Why a thistle?

But if beauty cannot be explained by necessity, then perhaps it is the product of chance. If so, beauty would be rare. But, on the contrary, nature abounds with beauty. As David Bohm declares: "Almost anything to be found in nature exhibits some kind of beauty both in immediate perception and in intellectual analysis." For instance, Portmann points out that virtually

all animals display some kind of symmetry.[29] And certain animals exhibit such
an astounding degree of symmetry that
they rival great works of art. (See Figure 4
for one example.) In fact, every level of investigation discloses new worlds of beauty
in nature. Thoreau, for instance, describes
the beauty of a field of grass:

"Standing on J.P. Brown's land, south
side, I observed his rich and luxuriant
uncut grass-lands northward, now waving
under the easterly wind. It is a beautiful
camilla, sweeping like waves of light and
shade over the whole breadth of his land,
like a low steam curling over it, imparting
wonderful life to the landscape, like the
light and shade of a changeable garment
. . . It is an interesting feature, easily over-
looked, and suggests that we are wading
and navigating at present in a sort of sea of
grass, which yields and undulates under
the wind like water."[30]

The simple elegance of a single clump of
grass is captured by the great Japanese
painters. A painting of the largest of the
grasses, bamboo, is seen in Figure 5.

The microscope reveals the hidden
geometry of cell structure in a single blade
of grass. Photographs of plant parts taken
through microscopes and through
scanning-electron microscopes are found in

FIGURE 4. Top and Side views of the *Discomedusa* illustrating octagonal symmetry.

FIGURE 5. Hsu Wei: Bamboo

art galleries and museums because of their stunning beauty.

Within the living cell, x-rays reveal the structure of the DNA molecule, the template of life, which James Watson, co-discoverer of its structure, calls beautiful. And finally, the atomic components of the DNA molecule itself are understood in terms of mathematical equations which possess an intellectual beauty, according to physicists.

Thus the poet, the painter, the biologist, the chemist, and the physicist all encounter the beauty of grass. Nature's beauty is not skin-deep. It penetrates the marrow. In all natural things, living and non-living, and at every level within each living thing, from grassy plain to electron, proton, and neutron, beauty saturates nature. Such abundant beauty of so many kinds and at so many levels could never come from chance. Physicist Henry Margenau concludes that nature's beauty is not reducible to either chance or necessity:

"We do not believe that beauty is only in the eye of the beholder. There are objective features underlying at least some experiences of beauty, such as the frequency ratios of the notes of a major chord, symmetry of geometric forms, or the aesthetic appeal of juxtaposed complementary colors. None of these has survival value, but

all are prevalent in nature in a measure
hardly compatible with chance. We marvel
at the song of the birds, the color scheme of
flowers (do insects have a sense of aesthetics?), of birds' feathers, and at the incom-
parable beauty of a fallen maple leaf, its
deep red coloring, its blue veins, and its
golden edges. Are these qualities useful for
survival when the leaf is about to decay?"[31]

If neither chance nor necessity can explain beauty, there must be something besides those alternatives. Whenever a cause
acts by necessity, there is a reason why it
acts as it does, but it is closed to all paths
but one. Chance, on the other hand, is open
to alternatives, but there is no reason why
one occurs rather than another. The middle ground between these two extremes is
a cause open to alternatives but with a
reason why one occurs rather than
another. Is there anything in our experience that operates in such a manner?
Clearly there is—our own minds.

Consider for a moment a craftsman fash-
ioning a bread knife for his own use. The
new knife will have a blade by necessity,
since it could not cut bread without one.
But the ornate, inlaid design of the handle
we cannot attribute to necessity, since a
knife can cut perfectly well with no decoration at all. The craftsman chooses freely to
embellish his work with ornament. He can
add the decoration or leave it out. And if he

adds it, he has an unlimited variety of designs to choose from. The knife's ornament is thus open to alternatives and yet has a reason for being there: the artist wants not only a useful knife but a beautiful one. The decoration is produced neither by chance, nor by necessity, but by an act of free choice. A mind choosing freely, then, is the middle ground between chance and necessity.

In the same way, since beauty is so abundant in nature it cannot arise from chance; there must be some reason for it. But that reason cannot be restricted to one path, since there is no absolute necessity that animals, plants, and non-living things exhibit beauty in the first place. Therefore, the beauty found in nature appears to proceed from a cause not bound by necessity and yet with a reason for acting. Such a cause is a mind. Therefore, a mind is responsible for the beauty of nature. That mind, standing behind nature, all men call God.

In Chapter III we saw how the New Story re-unites the sciences and the fine arts through the understanding of beauty. Poets, in their contemplation of nature's beauty, also recognize in it the work of a Mind. For instance, Thoreau, seeing that beauty cannot be explained by necessity, detected the Divine Artist behind nature:

"We are rained and snowed on with

gems. What a world we live in! Where are the jeweler's shops? There is nothing handsomer than a snowflake and dewdrop. I may say that the maker of the world exhausts his skill with each snowflake and dewdrop he sends down. We think that the one mechanically coheres and that the other simply flows together and falls, but in truth they are the product of *enthusiasm,* the children of an ecstasy, finished with the artist's utmost skill."[32]

One perceives the divine in the snowflake, in the sunset, in the field of grass; beauty's majesty and glory bear the unmistakable signature of God. "Beauty alone," says Thomas Mann, "is both divine and visible." And Emerson remarks: "Never lose an opportunity of seeing anything that is beautiful; for beauty is God's handwriting—a wayside sacrament. Welcome it in every fair face, in every fair sky, in every fair flower, and thank God for it as a cup of blessing."[33] Elizabeth Barrett Browning captures the same sentiment in two brief lines:

> God Himself is the best Poet,
> And the Real is His song.[34]

In the New Story, then, the origin of the universe, the structure of the universe, and the beauty of the universe, all lead to the same conclusion—God is.

V

MAN AND SOCIETY

If man is only a material being, as the Old Story maintains, then it makes sense to use simpler material things such as machines as models for human behavior. Every machine has a driving force that makes it function, such as steam, or electricity, or internal combustion. Thus the most fundamental thing about man in the psychology of the Old Story is his driving force. In the human machine material forces take the form of instincts and passions. They are the source of all human action. Mind cannot rule since it is a by-product of matter. The key to human psychology, then, is to discover the most powerful passion or instinct that drives man, overriding all else.

While psychologists of the Old Story agree that the instincts and passions drive man, they disagree on which is the most basic instinct. Some, like Hobbes, claim it is the fear of death; Malthus says it is hunger; Freud says it is the sex instinct. But the primal drive once assigned, the psychologies of the Old Story look similar.

Hobbes, for instance, formulated one of the earliest Old Story psychologies. In the 1630's he travelled to the Continent where he was introduced into scientific circles. He met Galileo in Florence and became a devotee of the new experimental sciences. Hobbes adopted methodological materialism and enthusiastically pursued the possibility of reaching a new scientific understanding of man by assuming matter alone.

Hobbes introduces his *Leviathan* by an explicit comparison of man to a machine. He compares the parts of human society to the parts of a machine as well. And in explaining human behavior, Hobbes assigns the fundamental role to the passions. He assigns differences of intelligence in men to differences in their passions: "The causes of this difference of wits, are in the passions." Consequently, a man without strong passions will lack drive: "he may be so far a good man, as to be free from giving offence; yet he cannot possibly have either a great fancy, or much judgment. . . . As to have no desire is to be dead: so to have weak passions is dullness."[1]

Hobbes describes the inevitable conflict that follows from man's passionate nature:

"Competition of riches, honour, command, or other power, inclines to contention, enmity and war: Because the way of one competitor, to the attaining of his desire, is to kill, subdue, supplant, or repel the other."[2]

If man lived in this natural condition without the restraints of government, there would be no place for the civilized things of life, says Hobbes; "No arts; no letters, no society; and which is worst of all, continual fear, and danger of violent death; and the life of man, solitary, poor, nasty, brutish and short."[3]

A grim picture. But man has invented the state to remedy this intolerable situation. Hobbes explains how self-preservation is the foundation of political society:

"The final cause, end, or design of men, (who naturally love liberty and dominion over others,) in the introduction of that restraint upon themselves, (in which we see them live in common-wealths,) is the foresight of their own preservation, and of a more contented life thereby; that is to say, of getting themselves out from that miserable condition of war, which is necessarily consequent (as hath been shown) to the natural passions of men."[4]

On this basis, Hobbes argues that the state must be totalitarian, since nothing short of absolute power will be sufficient to subdue men's passions.

Three hundred years after Hobbes, Freud begins his study of man with the assumption that only matter exists: psychoanalysts "are at bottom incorrigible mechanists and materialists." Freud considers this the only scientific approach to psychology. Hence Freud adopts the same general mechanical model of man set out by Hobbes. Freud indicates this when he describes himself as "a psychologist who has always insisted on what a minor part is played in human affairs by the intelligence as compared with the life of the instincts."[5]

Freud maintains the sex instinct is the driving force in man: "sexual love—has given us our most intense experience of an overwhelming sensation of pleasure and has thus furnished us with a pattern for our search for happiness."[6] But one soon discovers how this pleasure drive leads to conflicts:

"What decides the purpose of life is simply the programme of the pleasure principle. This principle dominates the operation of the mental apparatus from the start. There can be no doubt about its efficacy, and yet its programme is at loggerheads with the whole world, with the macrocosm as much as with the microcosm. There is no possibility at all of its being carried through; all the regulations of the universe run counter to it."[7]

According to Freud, the unavoidable frustration of the pleasure instinct forces us to pursue substitutes:

"Life, as we find it, is too hard for us; it brings too many pains, disappointments and impossible tasks. In order to bear it we cannot dispense with palliative measures .... There are perhaps three such measures: powerful deflections, which cause us to make light of our misery; substitute satisfactions, which diminish it; and intoxicating substances, which make us insensitive to it.... Scientific activity is a deflection of this kind .... The substitutive satisfaction as offered by art, are illusions in contrast with reality, but they are none the less psychically effective."[8]

These "displacements of libido," Freud continues, such as "an artist's joy in creating, ... or a scientist's in solving problems or discovering truths" may "seem 'finer and higher'. But their intensity is mild as compared with that derived from the sating of crude and primary instinctual impulses; it does not convulse our physical being." Thus "the mild narcosis induced in us by art ... can do no more than bring about a transient withdrawal from the pressure of vital needs, and it is not strong enough to make us forget real misery." When translated into clinical psychiatry this approach sometimes causes difficulties with the patient. For instance, psychiatrist Lawrence Hatterer reports that "Many an artist has left a psychiatrist's office enraged by interpretations that he writes because he is an injustice collector or a sadomasochist, acts because he is an exhibitionist, dances because he wants to seduce the audience sexually, or paints to overcome strict bowel training by free smearing."[9]

What is more, man's crude instinctual impulses conflict with the aims of society according to Freud. This follows by necessity if "every civilization must be built up on coercion and renunciation of instinct," and if "among these instinctual wishes are those of incest, cannibalism and lust for killing." Simply to maintain itself, every society must suppress such desires in man. Such suppression brings privation and frustration to the individual, according to Freud. That is why he declares, "Every individual is virtually an enemy of civilization, though civilization is supposed to be an object of universal human interest."[10]

As a consequence, man can never be happy within society since society contradicts the inclinations of his nature.

Freud, therefore, concludes: "If civilization imposes such great sacrifices not only on man's sexuality but on his aggressivity, we can understand better why it is hard for him to be happy in civilization. In fact, primitive man was better off in knowing no restrictions of instinct."[11] Far from being the remedy Hobbes thought it was, society in Freud's view exacerbates man's misery.

After Freud, there arose in the 1920's a further development in the psychological tradition of the Old Story: behaviorism. The founder of behaviorism, John B. Watson, earned a doctorate in psychology from the University of Chicago in 1903. After making extensive experiments with rats, birds, and monkeys, Watson sought to establish psychology as an absolutely objective branch of natural science, having no more need of what he called introspection than chemistry or physics. In human actions, only behavior is observable by means of objective, external experience. Therefore, according to Watson, only behavior, not consciousness, is the proper subject matter of psychology.[12]

Behaviorism does not try to derive mind from matter but asks rather why science should admit mind at all if it is never the source of any scientific explanation. Consequently, behaviorism asserts that man's body is the only human reality and that the "mind" and all of its trappings must be eliminated from science. In this sense behaviorism is more radically materialistic than Freudian psychology. The transition from Freud to behaviorism represents a progression in the eclipse of mind.

Within behaviorism everything previously attributed to mind must be either eliminated or redefined in terms of externally observable behavior. Watson mentions that the first step he had to take as a scientific psychologist was to drop from his "scientific vocabulary all subjective terms such as sensation, perception, image, desire, purpose, and even thinking and emotion as they were subjectively defined." And he adds, "The behaviorist recognizes no such things as mental traits, dispositions or tendencies." The aspects of mind that cannot be denied must be redefined in terms of behavior alone. Thus thinking becomes "subvocal talking." Watson declares, "Talking and thinking . . . when rightly understood, go far in breaking down the fiction that there is any such thing as 'mental life'."[13]

What picture of man results from behaviorism? Watson's methodological materialism is clear when he speaks of the causes of human action: "Man . . . as a corollary of the way he is put together and of the material out of which he is made . . . must act (until learning has reshaped him) as he does act."[14] Matter and the structures of matter are proposed as the only causes of human actions. Man, on this view, is an inert piece of matter that must be put into action by external forces.

The model appears to be taken from physics: a body at rest will remain at rest unless acted upon by an outside force. Rather than himself acting, man is in reality acted upon. This explains Watson's predilection for stimulus-response explanations: "The rule, or measuring rod, which the behaviorist puts in front of him always is: Can I describe this bit of behavior I see in terms of 'stimulus and response'?"[15] By stimulus, Watson means anything that causes a physiological change in the organism. By response, he means observable, preferably measurable, behavior. It is interesting to note that the Latin word *stimulus* first meant "goad".

Watson mentions two aims of behaviorism: "to predict human activity" and "the formulation of laws and principles whereby man's actions can be controlled by organized society."[16] If man is nothing more than a material entity, there is no reason to think he could not be programmed like a machine. Watson met vehement opposition to his theories from many fronts. He comments that behaviorism is not for the squeamish:

"People are willing to admit that they are animals but 'something else in addition'. It is this 'something else' that causes the trouble. In this 'something else' is bound up everything that is classed as religion, the life hereafter, morals, love of children, parents, country and the like. The raw fact that you, as a psychologist, if you are to remain scientific, must describe the behavior of man in no other terms than those you would use in describing the behavior of the ox you slaughter, drove and still drives many timid souls away from behaviorism."[17]

The "something else" of themselves that people wish so stubbornly to retain is the mind. Because of the criticism of his views, and for other reasons, Watson himself, in 1921, left the academic community for good and took up a business career in advertising.

We have seen in previous chapters how the New Story of science began with relativity and quantum mechanics which demonstrated the centrality of mind even in physics. Second, the New Story was supported by the great neuroscientists of this century who uncovered evidence of the mind's autonomy and irreducibility to matter (see Chapter II). A movement in contemporary psychology converges on the same conclusion, the primacy of mind. After the Second World War, many psychologists felt that the subordination of mind to instinct in psychoanalysis and the elimination of the mind in behaviorism led to the dehumanization of man in psychology. This they considered intolerable in a discipline dedicated to the service of mankind. Finally, in the 1950's, there coalesced a "third force" in psychology (the other two forces being psychoanalysis and behaviorism).

Psychologist Frank T. Severin describes the new movement: "They do not speak with a single voice, they do not constitute a separate school of thought, nor do they specialize in any specific content area. All that unites them is the common goal of humanizing psychology."[18] At a national meeting of the American Psychological Association in 1971, this new movement decided to call itself *humanistic psychology*. It is the psychology of the New Story of science. First, we will describe the aims and methods of this new psychology, then we shall examine what it has to say about man, about happiness, and about the relation of the individual to society.

Psychologist Irvin L. Child of Yale University explains the fundamental orientation of the new psychology: "Humanistic psychology is defined by its model of man, by its insistence that the body of scientific knowledge about man will develop most usefully if it is guided by a conception of man as he knows himself rather than by some nonhuman analogy."[19] What need do we have for a *model* of man if we ourselves are men? A model is a provisional parallel taken from something outside the subject itself. It is useful when the nature of a thing is remote or obscure. For example, Niels Bohr's planetary model of the atom is useful for certain purposes. We have no way of knowing what it feels like to be an atom, but we do have inside information about what it is like to be human. Child continues:

"Humanistic psychology, then, consists of all those currents of psychological thought in which man is viewed somewhat as he normally sees himself—as a person rather than only as an animal or a machine. Man is a conscious agent; that is the starting point. He experiences, he decides, he acts. If there are conditions under which man can usefully be looked at entirely from the outside, as responding to external stimulation with the regular predictability of a machine, a mechanical model may be useful for those conditions. But humanistic psychology starts with the presumption that such conditions are special cases, that to build the whole of psychology on them would mean an impoverishment of psychology, a restriction that would prevent its general application to the understanding of man."[20]

Man is a conscious agent; that is the

starting point. The primacy of mind is the core of humanistic psychology as it is the central theme of the New Story of science. Severin adds as an essential characteristic of humanistic psychology that "consciousness or awareness is the most basic psychological process." As for method, Severin continues, "behavioristic theories of science are based to a large extent upon nineteenth-century assumptions which are no longer considered valid. . . . By incorporating the new insights of physicists and philosophers, psychologists should be able to devise methodologies more in keeping with their unique subject matter."[21]

Psychologist Carl Rogers echoes the same sentiment:

"In an attempt to be ultrascientific, psychology has endeavored to walk in the footsteps of a Newtonian physics. Oppenheimer has expressed himself strongly on this, saying that the worst of all possible misunderstandings would be that psychology be influenced to model itself after a physics which is not there any more, which has been quite outdated. . . . I think there is quite general agreement that this is the path into which our logical-positivist behaviorism has led us."[22]

Furthermore, if man has the capacity of

choosing freely, then there is no need to reduce all human behavior to sub-human instinctual mechanisms, no need to assume that a healthy man's conscious motives are not the real causes of his actions. Psychologist Viktor Frankl warns:

"Unmasking is perfectly legitimate; but I would say that it must stop as soon as one confronts what is genuine, genuinely human, in man. If it does not stop there, the only thing that the 'unmasking psychologist' really unmasks is his own 'hidden motive'—namely, his unconscious need to debase and depreciate the humanness of man."[23]

In the New Story, man is not a bundle of reflexes, drives, or psychic mechanisms, nor is he the by-product of outside forces. The New Story seeks a human model for the study of man, without which we can never help those in need. "We cannot really help man in his predicament if we insist that our concept of man be patterned after the 'machine model' or after the 'rat model'," admonishes Frankl.[24] And psychologist Rollo May argues:

"If we are to study and understand man, we need a human model. That sounds like a truism, and it ought to be one; the amazing thing is that it is not a truism at all. I am continually impressed by the surprise registered by our scientific colleagues in other disciplines such as physics and biology when they find us taking our models not only from their sciences, but often from outmoded forms of their science they have already discarded."[25]

The outmoded form of science May refers to is the mechanism and materialism of the Old Story.

Man has the capacity to act for the sake of goals that he himself selects. But goals are based on values. Thus the psychology of the New Story includes the study of values. Rogers explains that the new psychology includes the inner riches of the person:

"In this world of inner meanings it can investigate all the issues which are meaningless for the behaviorist—purposes, goals, values, choice, perception of self, perceptions of others, the personal constructs with which we build our world, the responsibilities we accept or reject, the whole phenomenal world of the individual with its connective tissue of meaning."

Severin adds, "any science that imagines itself to be value-free is long outdated."[26] Neuroscientist Roger Sperry agrees:

"According to our new views of consciousness, ethical and moral values become a very legitimate part of the brain science. They're no longer conceived to be reducible to brain physiology. Instead, we now see that subjective values themselves exert powerful causal influence in brain function and behavior. They're universal determinants in all human decision making, and they're actually the most powerful causal control forces now shaping world events."[27]

This means that man's intellectual life, his moral life, and his spiritual life are all just as real as his biological life. If the mind has a life of its own, independent of matter, then the attempt to reduce art, religion, history, morality, politics, and human institutions to primitive instincts or biological necessities is a hopeless program and insistence on it would cut us off from a genuine understanding of man.

Modern man looks to psychology for guidance to what is good in human affairs. Rollo May declares in the name of psychologists:

"We are the representatives of modern science who are ordained to function in the realm of man's mind and spirit. . . . We are handed by society, whether we wish it or not, the requirement of producing answers to the ultimate questions of ethics and the spirit."[28]

One central question of ethics concerns the priority of certain values over others. We choose things because we think they are good for us. But are some things intrinsically better than others? Is there any way to distinguish a hierarchy among human goods?

If man is only matter, then it is to be expected that the Old Story will emphasize material, bodily goods. The pursuit of the arts and even the pursuit of science itself may be seen as pallid substitutes for "sating crude and primary instinctual impulses."[29] Psychologist Abraham Maslow criticizes those who reduce all human activities to drives and instincts:

"Because the lowest and most urgent needs are material, for example food, shelter, clothes, etc., they tend to generalize this to a chiefly materialistic psychology of motivation, forgetting that there are higher, non-material needs as well which are also 'basic'."[30]

But if man's mind is non-material, as the New Story maintains, then it has a life of its own independent of matter and there will be spiritual goods as well as material goods. By spiritual goods we mean moral, intellectual, and aesthetic values. Frankl states the approach of the New Story: "Man's . . . spiritual aspirations as well as his spiritual frustrations should be taken

at face value and should not be tran-
quilized or analyzed away."[31]

What are the spiritual goods of man?
They can be divided into two broad
categories: the goods of the intellect and
the goods of character. The first category
includes intellectual knowledge, not only
science, but also artistic skill, good judgment in practical matters, and wisdom.
The goods of character include all the
praiseworthy qualities of will such as
generosity, courage, and honesty.

The first thing we notice about these
spiritual goods is that they are acquired
only by free choice. As Frankl puts it:

"Values . . . do not drive a man; they do
not *push* him, but rather they *pull* him. . . .
There cannot exist in man any such thing
as a moral drive, or even a religious drive,
in the same manner . . . as basic instincts.
Man is never driven to moral behavior; in
each instance he decides to behave
morally."[32]

The material goods can come to us from
nature or by chance. A man may be born
with bodily strength or he may win riches
in a lottery. By contrast, one is a biologist
or an honest man not by nature or by
chance, but only by choice. The spiritual
goods have to be chosen. Frankl witnessed
the moral freedom of man when he was a

prisoner in several Nazi concentration camps during World War II. He recalls, "We watched . . . some of our comrades behave like swine while others behaved like saints. Man has both potentialities within himself; which one is actualized depends on decisions . . . not on conditions."[33]

The spiritual goods are the characteristically human goods. They are acquired only by choice and they can be lost only by choice. For example, no one can be involuntarily stripped of the goods of character. Someone can, by force, seize our belongings or attack us bodily. We can lose our material goods against our will. But no one can force us to be unjust or cowardly unless we consent. Others can treat us like animals, but whether we act like animals, even in a concentration camp, is ultimately up to us.

The same holds for the intellectual goods. When all else is stripped from us, these remain. May illustrates this with a remarkable case:

"Christopher Burney, a young British secret service officer, was dropped behind enemy lines during World War II and captured by the Germans. He was put in solitary confinement, without a book, pencil or sheet of paper, for eighteen months. In his six-by-six cell, Burney decided that each day he would review in his mind lesson after lesson he had studied in school and college. He worked through theorems in geometry, the thought of Spinoza and other philosophers, outlined in his head the literature he had read, and so on. In his book *Solitary Confinement* he demonstrates how the 'freedom of the mind,' as he calls it, kept him sane for the eighteen solitary months."[34]

Frankl testifies that the spiritual goods give more strength than the bodily goods. He recalls how fellow prisoners at Auschwitz with inner spiritual resources fared better than others with superior bodily strength:

"Sensitive people who were used to a rich intellectual life may have suffered much pain (they were often of a delicate constitution), but the damage to their inner selves was less. They were able to retreat from their terrible surroundings to a life of inner riches and spiritual freedom. Only in this way can one explain the apparent paradox that some prisoners of a less hardy make-up often seemed to survive camp life better than did those of a robust nature."[35]

That man can choose independently of his upbringing or history, that he can determine his own direction in life, not only confers upon him great responsibility, it elevates him to a dignity far above that of the material world.

The New Story psychology recognizes mind and will as the highest faculties in man. Thus, instead of the mind being a place of escape or illusion, as with Freud, it is the realm of reality and fulfillment. If mind and will distinguish man from animal, then pursuing the fine arts, the sciences, and the goods of character are man's highest activities.

Sperry describes the new view of mind:

"Mind and consciousness are put in the driver's seat, as it were: they give the orders, and they push and haul around the physiology and the physical and chemical processes as much as or more than the latter processes direct them. This scheme is one that puts mind back over matter, in a sense, not under or outside or beside it. . . . The causal potency of an idea, or an ideal, becomes just as real as that of a molecule, a cell, or a nerve impulse."[36]

The mind and will not only rule the body, they also rule, and when necessary overrule, the emotions. And with subordination of the emotions to reason, harmony and happiness become possible for man. Maslow, after extensive professional experience, describes the person who has harmonized all the parts of his nature, not by annihilating the emotions, but by directing them in accordance with the judgment of reason:

"In healthy people only is there a correlation between . . . delight in the experience, impulse to the experience, or wish for it, and 'basic need' for the experience (it's good for them in the long run). Only such people uniformly yearn for what is good for them and for others, and then are able wholeheartedly to enjoy it, and approve of it. For such people virtue is its own reward in the sense of being enjoyed in itself. They spontaneously tend to do right because that is what they *want* to do, what they *need* to do, what they enjoy, what they approve of doing, and what they will continue to enjoy."[37]

On the other hand, some persons choose habitually to indulge their passions even when they contradict what reason knows to be good. Inner discord characterizes such persons as Maslow explains:

"What he wants to do may be bad for him; even if he does it he may not enjoy it; even if he enjoys it, he may simultaneously disapprove of it, so that the enjoyment is itself poisoned or may disappear quickly. What he enjoys at first he may not enjoy later. His impulses, desires and enjoyments then become a poor guide to living. He must accordingly mistrust and fear the impulses and the enjoyments which lead him astray, and so he is caught in conflict, dissociation, indecision; in a word, he is caught in civil war."[38]

In the New Story, then, harmony within man's nature is possible. By the same token, harmony between the individual and society is also possible in the New Story. Conflict between men seems to follow from the Old Story's emphasis on material goods. Money and power, for example, do give rise to competition. But the spiritual goods do not. They are, by nature, common goods. They promote cooperation, since everyone can share in them without anyone's portion being diminished. The spiritual values of truth, beauty, and goodness illustrate the point.

Truth is a common good. Unlike material goods, it can be shared without loss. In *Gulag Archipelago,* Alexandr Solzhenitsyn describes a Soviet prison camp where a group of scientists and scholars, stripped of all bodily and external goods, are put to hard labor and given only a few ounces of bread a day:

"At the Samarka Camp in 1946 a group of intellectuals had reached the very brink of death: They were worn down by hunger, cold, and work beyond their powers. And they were even deprived of sleep. They had nowhere to lie down. Dugout barracks had not yet been built. Did they go and steal? Or squeal? Or whimper about their ruined lives? No! Foreseeing the approach of death in days rather than weeks, here is how they spent their last sleepless leisure, sitting up against the wall: Timofeyev-Ressovsky gathered them into a 'seminar', and they hastened to share with one another what one of them knew and the other did not—they delivered their last lectures to each other. Father Savely— spoke of 'unshameful death', a priest academician—about patristics, one of the Uniate fathers—about something in the area of dogmatics and canonical writings, an electrical engineer—on the principles of the energetics of the future, and a Leningrad economist—on how the effort to create principles of Soviet economics had failed for lack of new ideas. Timofeyev-Ressovsky himself talked about the principles of microphysics. From one session to the next, participants were missing—they were already in the morgue.

"That is the sort of person who can be interested in all this while already growing numb with approaching death—now that is an intellectual!"[39]

What a magnificent tribute to the life of the mind! Not only are the spiritual goods always accessible, they can even be shared when all else is taken away.

Beauty, too, is a common good. Even under conditions of extreme deprivation it can be shared without loss. Viktor Frankl recounts from personal experience in a Nazi concentration camp:

"In camp, too, a man might draw the attention of a comrade working next to him to a nice view of the setting sun shining through the tall trees of the Bavarian woods (as in the famous water color by Dürer), the same woods in which we had built an enormous, hidden munition plant. One evening, when we were already resting on the floor of our hut, dead tired, soup bowls in hand, a fellow prisoner rushed in and asked us to run out to the assembly grounds and see the wonderful sunset. Standing outside we saw sinister clouds glowing in the west and the whole sky alive with clouds of ever-changing shapes and colors, from steel blue to blood red. The desolate gray mud huts provided a sharp contrast, while the puddles on the muddy ground reflected the glowing sky."[40]

The same holds for goodness: goods of character are goods for all. The generous, courageous, honest man is a benefit not only to himself but to everyone around him. A good man is a public good. In *Man's Search for Meaning,* Frankl records many acts of heroism: "We who lived in concentration camps can remember the men who walked through the huts comforting others, giving away their last piece of bread."[41]

Because the spiritual goods promote cooperation rather than conflict among men, Maslow concludes, "We can now reject the almost universal mistake that the interests of the individual and of society are of *necessity* mutually exclusive and antagonistic, or that civilization is primarily a mechanism for controlling and policing human instinctoid impulses."[42]

VI

THE WORLD

The Old Story's picture of the world begins with its understanding of sense perception which it takes to be a material change. If we consider material changes, we notice that the same cause produces different effects. Apply the same amount of heat to 10 grams of wax, to 10 grams of water, and to 10 grams of gun powder. The wax melts, the water boils, and the gun powder explodes in a puff of smoke. These differences are not due to the cause, since it is identical in each case. The different results arise from the kind of matter in question and from its internal molecular structure. The melting, the boiling, and the exploding are facts about the wax, the water, and the gun powder, not about the heat source.

Now the Old Story of science applies this

material model to sense perception. If an external stimulus causes a change in a sense organ, that organ's response is dictated by the kind of matter the organ is made of and by the organ's structure. From this it follows that sense perception is primarily a fact about the sense organ itself and relates to the external cause only indirectly. On this view, we can never attribute any quality of our sensations to external things, any more than we attribute melting, boiling, or exploding to the flame that heats the wax, the water, and the gun powder. We might postulate an external object, a thing in itself, that causes our sensations, but nothing could ever be known about it since the whole content of sensation comes from the sense organ itself. This same reasoning holds for man's intellect and his other knowing faculties. Thus, if sensation is a material change, it appears that knowledge of the world is impossible.

This materialist understanding of man's knowing faculties and its inherent skepticism dominate the Old Story of science from its inception. The Old Story, however, only gradually comes to deny the possibility of knowing the world. Sir Francis Bacon, early formulator of the inductive method of science, begins by casting suspicion on the senses in general: "When the sense does apprehend a thing its apprehension is not much to be relied upon. For the testimony and information of the sense has reference always to man, not to the universe; and it is a great error to assert that the sense is the measure of things."[1]

Nor does the doubt stop with the senses. It encompasses even man's highest power, the human intellect:

"The intellect ... is far more prone to error than the sense is. For let men please themselves as they will in admiring and almost adoring the human mind, this is certain: that as an uneven mirror distorts the rays of objects according to its own figure and section, so the mind, when it receives impressions of objects through the sense, cannot be trusted to report them truly, but in forming its notions mixes up its own nature with the nature of things."[2]

Bacon implicitly assumes a materialist model of mind when he says the intellect "in forming its notions mixes up its own nature with the nature of things." But if a man can trust neither his senses nor his intellect, how can he ever be certain of anything? Bacon thinks he has the antidote: experiment. He hopes to correct the deceptions of the senses by such "experiments ... as are skillfully and artificially [i.e., artfully] devised for the express purpose of determining the point in question. To the immediate and proper perception of the sense, therefore, I do not give much weight; but I contrive that the office of the sense shall be only to judge of the experiment, and that the experiment itself shall judge of the thing." At this point one obvious question arises: if the senses cannot be trusted to report the world faithfully, how can they be trusted to report experiments? Bacon does not resolve this problem but insists that only by scientific experiment can man know the world. Bacon maintains that "the nature of things betrays itself more readily under vexations of art than in its natural freedom"[3] Since Bacon, the Old Story has affirmed the primacy of experiment over common experience.

Galileo considers mathematics to be the key feature of the scientific method, which alone can procure for man valid knowledge of the natural world. He says of the universe:

"We cannot understand it if we do not first learn the language and grasp the symbols, in which it is written. This book is written in the mathematical language, and the symbols are triangles, circles and other geometrical figures, without whose help it is impossible to comprehend a single word of it; without which one wanders in vain through a dark labyrinth."

Galileo sees mathematics as the new logic of science: "We do not learn to demonstrate from the manuals of logic, but from the books which are full of demonstrations, which are the mathematical and not the logical."[4]

He is confident that mathematics is the key to nature because of his many discoveries in mechanics where natural phenomena follow the principles of geometry. This fertility of mathematics in natural science suggests to Galileo that the world outside the mind is comprised only of mathematical properties. In a work called *The Assayer,* published in 1623, he explains what he understands by matter:

"Whenever I conceive any material or corporeal substance, I immediately feel the need to think of it as bounded, and as having this or that shape; as being large or small in relation to other things, and in some specific place at any given time; as being in motion or at rest; as touching or not touching some other body; and as being one in number, or few, or many. From these conditions I cannot separate such a substance by any stretch of the imagination."[5]

In a word, matter is mathematized. Only the quantitative properties, the measurable properties of matter, can be considered part of the real world: boundaries, shape, size, place, time, motion, contact, and number. And why is Galileo so confident that these qualities exist in external objects? It is because he cannot *conceive* of a body without them: "From these conditions I cannot separate such a substance by any stretch of the imagination."[6] This implies that our certitude concerning the reality of these properties comes from the way we must think about them, from the nature of our own minds.

And what of the non-mathematical qualities? Where do they come from? According to Galileo, these qualities come not from the world but from us: "Many sensations which are supposed to be qualities residing in external objects have no real existence save in us." Galileo explains: "I think that these tastes, odours, colours etc., on the side of the object in which they seem to exist, are nothing else than mere names, but hold their residence solely in the sensitive body; so that if the animal were removed, every such quality would be abolished and annihilated." He does not say the same for the mathematical qualities, however: "If ears, tongues, and noses were removed, shapes and numbers and motions would remain but, not odors or tastes or sounds."[7]

In previous chapters we have seen the Old Story's claim that for the purposes of science matter alone exists. Here we see that claim intensified—only the quantitative attributes of matter are real. Colors, odors, flavors, sounds—and, we might add, beauty and purpose—are not part of the real world.

According to the Old Story of science, then, there is not one world but two. Since Bacon and Galileo, the Old Story has assumed as a part of the method of science a divorce between the realms of the mind and matter. The parts of this dichotomy were eventually labelled the "subjective" and the "objective," (The modern use of those terms began in the seventeenth century.) From this division into two worlds there follow three consequences for the scientific method. *First,* in the Old Story, science is expected to study the "objective" world of matter rather than the "subjective" realm of mind. The goal of science is to describe the world without reference to mind. *Second,* doubt and skepticism become an integral part of the scientific method. And *third,* the scientific method alone, employing experiments and expressing itself in mathematical language, provides knowledge about the world.

These Old Story views have had decided effects on philosophy and the arts. Modern philosophy can be understood as the gradual unfolding of the assumption that sensing and understanding are material changes. Descartes, considered the father of modern philosophy, describes sense perception as if it were an activity of matter: "We have to think of the external shape of the sentient body as being really altered by the object precisely in the manner in which the shape of the surface of the wax is altered by the seal." This beginning leads Descartes to doubt his senses and his intellect just as Bacon did. What guarantee is there, Descartes asks, that any of my thoughts or perceptions represent anything outside of me? But then he considers that he cannot doubt that his own thoughts exist. Even if they represent nothing outside his mind, they are nevertheless real. Thinking itself is indubitable. Consequently, he enunciates his famous, "I think, therefore, I am." For Descartes the "I" is more certain than the world.[8]

Virtually all philosophers after Descartes adopt this subjective starting point, beginning with the thinking self rather than the world. This is taken as the natural beginning for philosophy even to this day. For example, Jean-Paul Sartre declares, "Subjectivity of the individual is indeed our point of departure, and this for strictly philosophic reasons. . . . There can be no other truth to take off from than this: *I think, therefore, I exist.* There we have the absolute truth of consciousness becoming aware of itself." Albert Camus agrees,

"Everything begins with consciousness and nothing is worth anything without it."[9]

According to Descartes, then, the first certitude is not the world but the thinking self. With this starting point, the existence of the world and of other minds becomes a serious problem for Descartes and for all of subsequent Western philosophy. Descartes uses two devices, his idea of God and his clear and distinct ideas, to deduce the existence of a world outside himself. Subsequent thinkers were not persuaded by either of these devices. As one historian put it, everyone was convinced of the existence of the external world until Descartes proved it. But regardless of whether Descartes' procedure is valid, the world he claims to regain at the end of his universal doubt is not the world of common experience he began with. It is the Galilean world of matter in which mind plays no role. For Descartes, matter itself is devoid of all sensible attributes except quantity, and the whole material universe, including Descartes' own body, is a machine acting out of mechanical necessity.

Despite his material determinism, Descartes tries to maintain that there is a God and that the human mind is nonmaterial. For this reason his philosophy, at first glance, might be confused with the New Story of science. But a closer inspection shows that Descartes also asserts the essential elements of the Old Story world view. As we saw in Chapters III and IV, he banishes beauty and purpose from natural science. He also adopts Galileo's division of sense qualities into those that are real and those that are in the mind only. The world Descartes ends up with after his skeptical doubts is the mathematical, mechanical world of Galileo, a world that leaves no room for free choice, beauty, purpose, or mind. Moreover, that world offers Descartes no evidence either for God or for a soul in man. All Descartes' arguments for these tenets are based on introspection, coming from within Descartes' mind alone. This is quite unlike the New Story of science which draws its evidence for beauty, free choice, purpose, God, and the immateriality of mind from the world itself: the Big Bang, the Anthropic Principle, modern neuroscience, and twentieth-century physics.

Descartes finds himself saddled with two incompatible worlds: the mechanistic, external world of matter and the unrelated, internal world of the mind. He never succeeds in reconciling them. Subsequent thinkers try to resolve the tension either by denying the immateriality of mind and affirming a complete materialism, or by denying the world and affirming a completely subjective mind. Thus modern philosophers tend to be either materialists or idealists.

After Descartes, the world of common experience is reduced further by philosopher George Berkeley who goes beyond Galileo and Descartes. In the name of defending religion from the challenge of Old Story science, Berkeley tries to discredit materialism by denying the existence of matter. One of his arguments is to attack the privileged status of the so-called objective qualities: "Let anyone consider those arguments which are thought manifestly to prove that colors and taste exist only in the mind, and he shall find they may with equal force be brought to prove the same thing of extension, figure and motion."[10] Berkeley even borrows some of his predecessors' arguments to substantiate this claim. For instance, Berkeley points out that any object, when brought closer to the eye, appears to grow larger and yet there is no real change in the object. Therefore, he concludes, size is not something absolute in the object but is only a perception existing in our minds.

Material substance becomes an unnecessary hypothesis, since it is no longer needed to support extension, motion, and the other mathematical qualities. And because all qualities are only ideas in our minds, Berkeley concludes, the only kind of substance is the self or mind in which all these ideas exist. Thus Berkeley admits only spiritual substance: "Nothing properly but persons, i.e., conscious things, do exist; and other things are not so much existence as manner of the existence of persons."[11]

In this way, Berkeley advances the eclipse of the world begun by Galileo and Descartes. In Berkeley's philosophy matter itself vanishes from the world; only spiritual substance remains. The next, almost inevitable step is taken by David Hume who assumes a complete phenomenalism and denies both material and spiritual substance. For Hume nothing remains but the subjective perceptions themselves. In addition, he launches a now famous skeptical attack on man's knowledge of cause and effect, reducing it to custom.[12]

The eclipse of the world is complete; Matter cannot be known. But if so, science itself becomes impossible. If we know nothing but our own perceptions and ideas, how can we do physics? How could Newtonian physics predict phenomena? Immanuel Kant sets out to solve that very problem. Conceding to Hume that human experience is nothing more than a series of perceptions or appearances, Kant makes an ingenious attempt to ground the necessity of scientific laws not in nature but in the structure of the human mind. The laws of natural science have universality, says Kant, because man simply cannot think of natural things in any other way: "The understanding does not derive its laws . . . from, but prescribes them to, nature."[13] A hint of this approach is found already in Galileo when he argues that mathematical properties must exist in matter because he cannot *conceive* of any body without them.

It is the philosopher's task, according to Kant, to identify and describe the various necessary categories of the mind and the *a priori* forms that sensation imposes upon objects. In attempting to do this, Kant declares that Euclidean space and all the categories of Newtonian physics are the inevitable way the human mind—because of its very structure—must understand the material world. Note that deriving the universality and necessity of science from the *structure* of the mind presupposes the materialistic model of mind that we saw at the beginning of this chapter.

Kant holds that freedom of the will, immortality, and God, though necessary postulates for ethics, cannot be proven and that the material world offers no evidence for these beliefs. This paradox arises because Kant's philosophy incorporates two worlds, not one. First, there is the internal world of the mind in which we seem to act freely. Second, there is the world of external phenomena that obey the necessary laws of Newton's physics. Like Descartes, Kant feels compelled to posit two irreconcilable worlds.

Subsequently, Georg Hegel tries to overcome the tension and disunity of Kant's two worlds by assuming that the world of external phenomena is a projection of the mind itself. Hegel attempts to construct a completely mentalistic system in which all art, religion, science, all human history, and the whole of nature—including the "thing in itself"—are but creations of the collective human mind, or "Absolute Spirit" as Hegel calls it. Where Bacon says that the mind "in forming its notions mixes up its own nature with the nature of things,"[14] Hegel says the mind's notions *are* the nature of things. With Hegel, the collective human consciousness itself is the cause, source, and explanation of all that exists. The mind creates reality, it creates the world, it creates truth.

Hegel claims to achieve a grand unity of world and mind. But at what cost? What is the value of a "truth" that is a pure mental fabrication? It tells us nothing about the world outside of our own thoughts. In reaction to Hegel's system, contemporary philosophy is suspicious of all speculative thought, considering it mere system-building, the human mind spinning out its own arbitrary creations. It is generally assumed to be impossible to reach any truth about the world. Sartre writes, "Outside the Cartesian *cogito,* all views are only probable." Or as Albert Camus puts it, "All thought is anthropomorphic. . . . The mind that aims to understand reality can consider itself satisfied only by reducing it to terms of thought." All thought is distortion or invention. Camus implicitly assumes, as do all his predecessors back to Bacon and Galileo, that the senses and the mind are acted upon in the same way that matter is acted upon. From that it follows that we can know only our own perceptions and never things in themselves. All thought is anthropomorphic.[15]

The atheistic existentialists stress the meaninglessness of the world and its unintelligibility. Camus says, "That denseness and that strangeness of the world is the absurd." He continues, "And what constitutes the basis of that conflict, of that break between the world and my mind, but the awareness of it?" If man had no mind, he could at least enjoy an animal happiness. But cursed with a mind, man is the only creature who searches the world for a meaning that is not there. What good is a mind if it cannot know anything? Camus protests:

"If I were a tree among trees, a cat among animals, this life would have meaning, or rather the problem would not arise, for I should belong to this world . . . This world to which I am now opposed by my whole consciousness . . . This ridiculous reason is what sets me in opposition to all creation."[16]

Contemporary philosophy is marked by a radical intellectual despair, the abandonment of all hope of knowing the world. Camus observes, "With the exception of professional rationalists, today people despair of true knowledge."[17] The same message comes from many sides: reason is impotent.

Philosopher Friedrich Nietzsche complains that all modern philosophy is "reduced to theory of knowledge . . . a philosophy which cannot get past its own threshold and has painstakingly *forbidden* its own right to enter—this surely is philosophy in its last throes, an end, an agony, something that arouses compassion."[18] Nietzsche's own proposal is to abandon the very notion of truth and replace it with the will to power.

In contemporary philosophy, then, the eclipse of the world is complete and absolute. The tiny, seemingly innocent seed planted by Bacon and Galileo in the seventeenth century has taken more than 300 years to produce in our day the bitter fruit of intellectual despair, nihilism, and the alienation of man from the world. Such is the final outcome of philosophy within the confines of the Old Story. It follows that if nothing can be known about the world, then science itself is also lost. It, too, can be no more than another anthropomorphic system of constructs. Thus, paradoxically, the philosophic consequences of the Old Story undermine the Old Story itself.

The conflict in the Old Story between the subjective world of mind and the objective world of matter permeates not only modern philosophy but also modern art.

German expressionist painter Franz Marc, in describing the goal for the art of the future, manifests the artistic problems raised by the Old Story's two worlds: "Art will free itself from the needs and desires of men. We will no longer paint a forest or a house as we please or as they seem to us, but *as they really are*."[19] But how can an artist know an object otherwise than as it appears to his eye or to his imagination or to his mind? The problem of the "thing in itself" plagues the modern artist as well as the philosopher.

Painter Piet Mondrian champions non-figurative art because it "shows that 'art' is not the expression of the appearance of reality such as we see it, nor of the life which we live, but . . . is the expression of true reality and true life . . . indefinable but realizable in plastics." Man must be driven out of the world of art, just as he is excluded from the "objective" world of science. Painter and critic Edward Wadsworth sums up the artistic principle of objectivity: "A painting is no longer a window out of which one sees an attractive little bit of nature; nor is it a means of demonstrating the personal sentiments of the artist: it is itself, it is an object . . . the painter does not paint what he sees but what he knows *is*."[20]

Modern literature has also felt the demands of objectivity. This is especially true of those writers who consciously imitate the methods of science. The ideal of some is to become like Newton's detached observer, carefully avoiding all judgments of good and bad. Gustave Flaubert elaborates:

"I do not even think that the novelist ought to express his own opinion on the things of this world . . . I limit myself, then, to declaring things as they appear to me, to expressing what seems to me to be true. . . . Art should be raised above personal affections and nervous susceptibilities. It is time to give it the precision of the physical sciences by means of a pitiless method."[21]

Anton Chekhov, reflecting the intellectual despair of the Old Story, argues that a writer should eschew all judgments, much like a newspaper reporter or a witness in a court of law:

"It is not for the writers of fiction to solve such questions as that of God, of pessimism, etc. The writer's business is simply to describe who has been speaking about God or about pessimism, how, and in what circumstances. The artist must not be the judge of his characters and of their conversations, but merely an impartial witness. I have heard a desultory conversation of two Russians about pessimism—a conversation which settles nothing—and I must report that conversation as I heard it; it is for the jury, that is, for the readers, to decide on the value of it. My business is merely to be talented, i.e. to know how to distinguish important statements from unimportant, how to throw light on the characters, and to speak their language. . . . It is time that writers, especially those who are artists, recognized that there is no making out anything in this world."[22]

The art of the Old Story, no less than the philosophy, is entrapped by the objective-subjective dilemma. To be "objective" it seems a work must be devoid of values and content, becoming a clinical narrative. On the other hand, if a work is "subjective", it does not rise above idiosyncratic temperament and "nervous susceptibilities". On this view, Shakespeare becomes little more than the reaction of a particular temperament to Elizabethan

England; Mozart and Michelangelo, mere expressions of their times. To compound matters, an audience cannot view Shakespeare except through the particular lens of its age. The situation appears hopeless. Anatole France complains:

"There is no such thing as objective criticism any more than there is objective art, and all who flatter themselves that they put aught but themselves into their work are dupes of the most fallacious illusion. The truth is that one never gets out of oneself. . . . We are locked into our own persons as into a lasting prison. The best we can do, it seems to me, is gracefully to recognize this terrible situation and to admit that we speak of ourselves every time we have not the strength to be silent. To be quite frank, the critic ought to say: 'Gentlemen, I am going to talk about myself on the subject of Shakespeare, or Racine, or Pascal, or Goethe.'"[23]

Truth has vanished. Only viewpoints remain.

Before we turn to the New Story, let us briefly recapitulate what we have seen so far. The Old Story's assumption that sensation is a material change implies the impossibility of knowing the world. To save science, Galileo splits the world into a "subjective" realm of mind and an "objective," mathematical world of matter. Hence, to know the world, science must rely not on common experience but on specialized experiments and mathematics. Modern philosophy, however, fully unfolds the implications of the materialist theory of sensation and understanding, showing that such an approach leads to the progressive eclipse of the world, to eventual intellectual despair, to the absurdity of man's having a mind, and to the undermining of science itself. Finally, art in the Old Story is also hamstrung by the objective-subjective dilemma, unable to incorporate values, unable to define itself apart from the projection of temperament.

In contrast to the Old Story, twentieth-century neuroscience asserts that sense perception is not a material change, not an activity of matter. (See Chapter II.) Sherrington, Eccles, and Penfield agree that all the physical changes in the sense organ, in the nerve paths and in the brain take us only to the threshold of sensation. Sense perception, though requiring material change, is itself non-material. This means that sensing is an act of pure passivity, pure receiving. The sense power does not mix its own nature with the nature of things as Bacon thought, because the sense power has no material nature. The same holds even more so for the human intellect, which apparently has no bodily organ.

If our senses add nothing to the object sensed, then what they report must be in the world. This means that there is only one world, a world that our senses make us genuinely know. As Schrödinger writes: "The world is given to me only once, not one existing and one perceived. Subject and object are only one. The barrier between them cannot be said to have been broken down as a result of recent experience in the physical sciences, for this barrier does not exist."[24] The discoveries of twentieth-century neuroscience have brought us to recognize that this barrier never existed.

When a man steps onto a scale that registers "187," that "187" is a fact about both the man and the scale. In a similar way knowledge is a fact about both the knower and the object known. The division into two worlds, one "objective" and the other "subjective" is neither natural nor inevitable. And it is certainly not a part of the scientific method. It is not needed to discover any truth about the world—in fact, if taken seriously, it renders knowledge of the world quite impossible. The division into "subjective" and "objective" worlds is necessary only within the materialism of the Old Story of science. And since the New Story makes no assumption of materialism, it has no need to dismiss our perceptions as "subjective," no need to split the world in two. "This objective-subjective distinction is illusory," concludes Eccles.[25]

If the objective-subjective dichotomy is a false division of experience, then what division does the New Story offer in its stead? Werner Heisenberg provides the answer:

"Since the time of Galileo the fundamental method of natural science has been the experiment. This method made it possible to pass from general experience to specific experience, to single out characteristic events in nature from which its 'laws' could be studied more directly than from general experience."[26]

Experience is properly divided into common experience and specialized experience. Common experience suffices, for instance, to tell us that heavy objects fall, but specialized experience (experiments, measurements, and mathematical calculations) is needed if we wish to know how fast heavy objects fall or whether their fall is a uniform motion or an accelerating motion.

Mathematics does not banish ordinary thought from physics. The first conception of a physical theory is non-mathematical. Einstein remarks:

"Fundamental ideas play the most essential role in forming physical theory.

Books on physics are full of complicated mathematical formulae. But thoughts and ideas, not formulae, are the beginning of every physical theory. The ideas must later take the mathematical form of a quantitative theory, to make possible the comparison with experiment."[27]

And even in expressing his discoveries, the scientist cannot rely exclusively on mathematics or on the technical vocabulary of his specialty. He must graft his branch of discovery onto the trunk of common experience that expresses itself in ordinary language. Heisenberg insists that "Even for the physicist the description in plain language will be a criterion of the degree of understanding that has been reached." And Niels Bohr adds, "All account of physical experience is, of course, ultimately based on common language, adapted to orientation in our surroundings and to tracing relationships between cause and effect."[28]

Now the Old Story of science rejects common experience, considering it unreliable. Bacon, Galileo, and Descartes try to replace it with the specialized experience of science. The scientists of the New Story, however, show great regard for common experience, pointing out that science does not replace it but rather builds on it as on a foundation. Heisenberg, for example, declares:

"One of the most important features of the development and the analysis of modern physics is the experience that the concepts of natural language, vaguely defined as they are, seem to be more stable in the expansion of knowledge than the precise terms of scientific language, derived as an idealization from only limited groups of phenomena. This is in fact not surprising since the concepts of natural language are formed by the immediate connection with reality; they represent reality."[29]

Common experience is immediately connected with reality, with the world. With common experience no theory, no instrument or any other middle man intervenes between us and the world. Common experience has the immediacy of the sense of touch which is always right on top of what it apprehends. Heisenberg continues the same point:

"Scientific concepts are idealizations; they are derived from experience obtained by refined experimental tools, and are precisely defined . . . But through this process of idealization and definition the immediate connection with reality is lost. The concepts still correspond very closely to reality in that part of nature which had been the object of research. But the correspondence may be lost in other parts containing other groups of phenomena."[30]

The New Story's restoration of common experience gives the world back to man. Not the Galilean or Cartesian world, stripped of most sense qualities and devoid of mind, nor any of the narrower subjective worlds of Berkeley, or Hume, or Kant, or Sartre, but the world we live in, with all its fullness and richness. We have already seen in Chapters II, III, and V how the New Story supports what common experience has always said about our capacity to choose freely, about the beauty of nature, and about what man is. Sperry observes that the world was taken as

"a matter of common sense until science came along and began telling us otherwise. Ever since, there's been a growing conflict of culture and world view between scientists and the rest of society, felt most keenly in the humanities and especially in those disciplines most concerned with moral values. Perhaps what I'm saying here, in effect, is an admission: The humanities and common sense were on the right track all along, and we in science were misled."[31]

Heisenberg adds that acknowledging common experience entails a different world view from the Old Story:

"Keeping in mind the intrinsic stability of the concepts of natural language in the process of scientific development, one sees that—after the experience of modern physics—our attitude towards concepts like mind or the human soul or life or God will be different from that of the nineteenth century, because those concepts belong to the natural language and have therefore immediate connection with reality."[32]

A further consequence is that science is founded on certitude, not doubt. For if ever there is a doubt about a particular theory, one can always fall back on a more general, more stable truth. Doubt is not a principle of discovery of science. It is wonder, not doubt, that animates science. Erwin Schrödinger observes, "Curiosity is the stimulus. The first requirement of the scientist is to be curious. He must be capable of being astonished and eager to find out."[33] The capacity for wonder distinguishes the great scientist. Physicist Leopold Infeld, collaborator with Einstein, testifies to Einstein's profound capacity for wonder:

"From the time Einstein was fifteen or

sixteen years old (so he has often told me) he puzzled over the question: what will happen if a man tries to catch a light ray? For years he thought about this very problem. Its solution led to relativity theory. We see in this one example some important features of Einstein's genius. First, and above all, there is the capacity for wonder."[34]

The Old Story considers power or utility to be the goal of science. Hence Bacon's famous equation of knowledge with power. The Old Story sees science as either trying to conquer and control a hostile nature, or as building a material utopia on earth through ingenious scientific inventions. But if, according to the New Story, science is fired by wonder, then its goal is not power, but contemplation of the truth. Infeld comments:

"It is wrong to assume that the entire development of science is of a utilitarian character. It is not. Many of our speculations about atoms and about our universe were created because of man's curiosity, because of his desire to penetrate . . . deeper and deeper into the unknown. The utilitarian value of many of our theories may be nil, but they help us to comprehend the world we live in."[35]

The restoration of common experience in the New Story brings another benefit. It makes possible a new unity between the specialist and the common man. Intensive specialization brings difficulties in all the disciplines. Schrödinger comments, "Specialization is not a virtue but an unavoidable evil." It narrows the mind if disconnected from a more general understanding. Mathematician Morris Kline remarks, "The price of specialization is sterility. It may well call for virtuosity but rarely does it offer significance." Schrödinger lays it down as a principle that "The isolated knowledge obtained by a group of specialists in a narrow field has no value whatsoever, but only in its synthesis with all the rest of knowledge." The knowledge that arises from common experience furnishes the common ground on which all the specialists in science, in philosophy, and in the arts can meet. If one tries to replace common experience with specialized knowledge as the Old Story proposes, it would be difficult to find any ground of communication among specialists, or between the specialist and the non-expert. Schrödinger insists that in any science or discipline "If you cannot—in the long run—tell everyone what you have been doing, your doing has been worthless."[36]

As for philosophy, the New Story of science remedies the impoverishment of experience that characterizes philosophy since Descartes. Carl von Weizsacker comments: "The Cartesian subject is solitary. Its natural ties with reality are severed. . . . Only an 'I' of this kind can conceive the idea that his own thinking is the only immediate certainty." According to the New Story, there is no reason why the world of science and the world of the philosopher cannot be the same as the world we all live in. Viktor Frankl calls for a recognition of the certitudes of "ordinary living unbiased by theories." On those certitudes of common experience can be built a philosophy full of wisdom and worthy of man.[37]

Finally, the New Story's affirmation of common experience does not diminish the arts but enriches them and humanizes them. In the arts, the greatest works have always respected the continuity between common experience and specialized experience. Mozart describes one of his compositions in a letter to a friend: "There are passages here and there from which the connoisseurs alone can derive satisfaction; but these passages are written in such a way that the less learned cannot fail to be pleased, even without knowing why."[38] Every classic offers something to all levels of experience. Shakespeare's plays, lofty enough to nourish the most sophisticated mind, are yet concrete enough to entertain.

By founding his work on common experience, an artist can produce what is timeless and universal. No classic work of art is the mere expression of a temperament or the simple reflection of its age. Composer Roger Sessions remarks, "Bach and Mozart and Beethoven did not *reflect* Germany, they helped to create it. . . . So what we most ask of our composers is not 'American' music, but . . . music that is . . . deeply and completely conceived, the product of a mature vision of life."[39]

In the New Story of science, the ordinary man, the scientist, and the philosopher can know the world, and the artist can render the fullness and the richness of that world in his art. Goethe intended the following advice for the poet but it is equally apt for any artist, scientist, or philosopher:

"[A poet] deserves not the name while he only speaks out his few subjective feelings; but as soon as he can appropriate to himself and express the world, he is a poet. Then he is inexhaustible, and can be always new, while a subjective nature has soon talked out his little internal material, and is at last ruined by mannerism. People always talk of the ancients; but what does that mean, except that it says, turn your attention to the real world, and try to express it, for that is what the ancients did when they were alive."[40]

VII

THE PAST

What attitude does the Old Story adopt toward its predecessors? The development of modern experimental science in the late Renaissance initiated the Old Story; hence, its predecessors are the thinkers of the ancient world and the thinkers of the Middle Ages. According to the Old Story, these thinkers can contribute little if anything to our knowledge of nature because they all lived before the advent of experiments, sophisticated instruments, and modern mathematics. This means that modern science cannot be built on ancient foundations. Bacon articulates the Old Story approach: "It is idle to expect any great advancement in science from the ... engrafting of new things upon old. We must begin anew from the very foundations."[1]

The temptation in the Old Story is to think of modern science as replacing all else. Scientific progress is thought to come more by rupture than by continuity. In 1903 physicist Ernst Mach took a strong position reminiscent of Bacon's: "Our culture has gradually acquired full independence, soaring far above that of antiquity. It is following an entirely *new* trend. It centers around mathematical and scientific enlightenment. The traces of ancient ideas, still lingering in philosophy, jurisprudence, art and science constitute impediments rather than assets, and will come to be untenable in the long run in face of the development of our own views."[2] According to Mach the pre-scientific past is primarily a repository of error and unfounded conjecture. A purely historical interest in the past is unobjectionable, but one should not consider antiquity as a source of enlightenment.

The same attitude toward the past is found in the philosophy of the Old Story. Descartes sets the precedent when he grants that a study of the ancients may be useful for certain purposes but warns, "There is great danger lest in too-absorbed study of those works we should become infected with their errors." In developing his own philosophy, Descartes puts aside his predecessors and relies upon himself alone. After much study and travel he writes, "I was . . . unable to decide on any one person whose opinion seemed worthy of preference, and so had no option save to look to myself for guidance."[3]

The Old Story encourages the artist also to reject his predecessors and abandon continuity with the past. Reliance on previous artists is regarded as an impediment to originality and free creativity. In 1910 Umberto Boccioni drew up a manifesto of Futurist painters. It expresses well the attitude:

"We want to fight relentlessly against the fanatical, irresponsible, and snobbish religion of the past, which is nourished by the baneful existence of museums. We rebel against the groveling admiration for old canvases, old statues, old objects and against the enthusiasm for everything moth-eaten, dirty, time-worn, and we regard as unjust and criminal the usual disdain for everything young, new, and pulsating with life. . . .

"By our enthusiastic adherence to futurism we propose: to destroy the cult of the past, the obsession with the antique, the pedantry and formalism of the academies; to despise utterly every form of imitation; to extol every form of originality, however audacious, however violent; . . . to sweep from the field of art all motifs and subjects that have already been exploited; to render and glorify the life of today, unceasingly and violently transformed by victorious science."[4]

The Old Story's orientation to the past in science, in philosophy, and in the arts, then, is clear. What is the New Story's attitude toward the past? One can begin with a look at the New Story's treatment of the Old Story.

First of all, there is no wholesale rejection of the Old Story. The New Story retains as permanently valid all truths about matter discovered under the materialistic program. As Heisenberg notes, "Modern physics has changed nothing in the great classical disciplines of, for instance, mechanics, optics, and heat."[5] The New Story also retains all the valid principles of method developed by the pioneers of the Old Story, including the necessity of experiments, instruments, and higher mathematics to investigate nature.

The only part of the Old Story that the New Story does not assimilate is its materialism. From Bacon the New Story takes the tool of experiment, but not his radical skepticism about man's faculties stemming from his materialistic model of sensation and understanding. From Galileo the New Story retains the centrality of mathematics for progress in natural science, but not his claim that only mathematical qualities are real.

The New Story acknowledges all the genuine discoveries made under the aegis of the materialistic program, but it does not accept the claim that those discoveries validate materialism. Eccles illustrates the attitude of the New Story:

"Of course, I fully support scientific investigations on behaviour and conditioned reflexes and, in fact, all the present scientific programs of behaviouristic psychology. Furthermore, I agree that much of human behaviour can be satisfactorily explained on the basis of concepts developed in relation to these experiments. However, I differ radically from the behaviourists in that they claim to give a *complete* account of the behaviour of man, whereas I know that it . . . does not explain me to myself, for it ignores or relegates to a meaningless role my conscious experiences, and to me these constitute the primary reality—as doubtless it does to each one of you, my readers."[6]

Carl Rogers takes the same stance toward the psychologies of the Old Story:

"Each current in psychology has its own implicit philosophy of man. Though not often stated explicitly, these philosophies exert their influence in many significant and subtle ways. For the behaviorist, man is a machine, a complicated but nonetheless understandable machine, which we can learn to manipulate with greater and greater skill until he thinks the thoughts, moves in the directions, and behaves in the ways selected for him. For the Freudian, man is an irrational being, irrevocably in the grip of his past and of the product of that past, his unconscious.

"It is not necessary to deny that there is truth in each of these formulations in order to recognize that there is another perspective."[7]

In the New Story of science innovation does not mean iconoclasm. To build a new theory does not entail completely razing the previous one. Even scientific revolutions maintain a continuity with the past. For example, discussing his own theory of gravitation, Einstein compares it to Newton's: "No one must think that Newton's great creation can be overthrown in any real sense by this or by any other theory. His clear and wide ideas will forever retain their significance as the foundation on which our modern conceptions of physics have been built."[8]

The word revolution may be misleading. Relativity overthrew Newtonian physics as a world system. But Newton's inverse square law is still quite adequate to account for ordinary experience and is in fact deducible from general relativity as an approximation. Leopold Infeld states: "It is not quite correct to say that Einstein has proven Newtonian mechanics to be inapplicable. It is more correct to say that he has shown its limitations. But the region in which it works is still vast."[9] It is necessary to introduce the precision of relativity theory only when speaking of special cases, such as speeds approaching the speed of light. Newton's equations were perfectly adequate to put men on the moon and bring them home safely.

Every advance in science must preserve the truths of the past and build on them. Physicist Louis de Broglie provides the formula of permanence: "Any time a law has been incontestably verified for a certain degree of approximation (every verification always involves a degree of approximation) the result is established definitively and no further theorizing can reverse it. If this were not so, no knowledge would be possible."[10] A new theory may offer more precision than an older one but it never reverses anything genuinely verified. Because we are dazzled by the drama of what is novel in a new theory, we sometimes lose sight of the underlying continuity. Failing to see that continuity, one might imagine that physics is in constant upheaval with no foundations at all.

Not only does the New Story establish a new continuity with the scientific past, it even provides a new harmony with the pre-scientific era. In fact, the major principles of the New Story agree with the world view of the Middle Ages and ancient times. The world view that prevailed before Galileo and Bacon was Aristotelian physics. Therefore, to represent that world view we shall refer principally to Aristotle, citing other authors only by way of supplement.

Like the New Story of science, Aristotle affirms that mind is not reducible to matter. Like Sherrington and Eccles, he defines sense perception as a non-material activity: "By sense is meant what has the power of receiving into itself the sensible forms of things without the matter."[11] Aristotle adds that understanding is not an activity of matter, maintaining like Penfield that the human intellect has no bodily organ.[12] It is a matter of scholarly debate whether Aristotle maintained immortality of the individual. He does say, however, "Mind . . . alone is immortal and eternal."[13] And because he does not use a materialistic model to explain knowledge, Aristotle does not fall into the subjective-objective difficulties that plague philosophers from Descartes onward.

Consistent with his recognition of the immateriality of mind, Aristotle also affirms that man has the capacity to choose freely. He clearly distinguishes the will from the emotions and maintains that the will is a rational appetite. He defines choice as a decision resulting from the deliberation of reason: "Choice will be deliberate desire of things in our power."[14]

Aristotle holds further that the goods of character are acquired only by free choice: "External goods come of themselves, and chance is the author of them, but no one is just or temperate by or through chance."[15] And there are many passages in his treatises on ethics and politics that offer reasons for the superiority of the spiritual goods over the material goods.[16] He praises above all other human activities the life of the mind and the moral life where reason governs a man's actions and passions.[17]

As for beauty, Aristotle declares it to be found in the whole of nature but especially in living things. In his *Parts of Animals* he encourages the beginner to

"study every kind of animal without distaste; for each and all will reveal to us something natural and something beautiful. Absence of haphazard and conduciveness of everything to an end are to be found in nature's works in the highest degree, and the resultant end of her generations and combinations is a form of the beautiful."

Aristotle argues from nature's simplicity.

In his *Physics,* for example, he criticizes Anaxagoras who assumed an unlimited number of natural principles: "It is better to assume a smaller and finite number of principles, as Empedocles does."[18]

Moreover, the three elements of beauty discussed in Chapter III are found in the writings of Plato, Aristotle, Plotinus, and other thinkers of the pre-scientific era. For example, Thomas Aquinas writes, "Three items are required for beauty: first, integrity or perfection, for things that are lessened are ugly by this very fact; second, due proportion or harmony; and third, brilliance, thus, things that have a bright color are said to be beautiful."[19] The same three elements developed by Einstein and Gell-Mann 700 years later!

Aristotle also asserts, "Nature is a cause, a cause that operates for a purpose."[20] He sees a close alliance between nature's simplicity, its beauty, and its purposefulness: "Nature never makes anything without a purpose and never leaves out what is necessary."[21] Nothing lacking, nothing superfluous. Aristotle in many of his works insists that purpose is indispensable for a proper understanding of natural things, especially living things. And not only are individual organs to be understood and studied in light of what they are built for, the parts of the universe as a whole are ordered to a common end:

"We must consider also in which of two ways the nature of the universe contains the good and the highest good, whether as something separate and by itself, or as the order of the parts. Probably in both ways, as an army does; for its good is found both in its order and in its leader, and more in the latter; for he does not depend on the order but it depends on him. And all things are ordered together somehow, but not all alike—both fishes and fowls and plants; and the world is not such that one thing has nothing to do with another, but they are connected."[22]

Another principle of the New Story is observership as seen in relativity and quantum physics. Man's unique role as observer is recognized by Aristotle and by many subsequent pre-scientific thinkers. Aristotle opens his *Metaphysics* with the famous line, "All men by nature desire to know." He sees in animals no evidence of either aesthetic or intellectual contemplation of nature: "Animals other than man live by appearances and memories, and have but little of connected experience." They use their senses for utility only, never merely to acquire knowledge or to appreciate beauty:

"Dogs do not delight in the scent of hares, but in the eating of them, but the scent told them the hares were there; nor does the lion delight in the lowing of the ox, but in eating it; but he perceived by the lowing that it was near, and therefore appears to delight in the lowing."[23]

The principle of observership is taken up by many other ancient thinkers. Cicero writes "Man himself, however, came into existence for the purpose of contemplating . . . the world." Epictetus writes "God introduced man to be a spectator of God and of his works; and not only a spectator of them, but an interpreter." Saint Augustine observes that the world would be incomplete without man the observer: "Material things . . . help to make the pattern of this visible world so beautiful. It is as though, in compensation for their own incapacity to know, they wanted to become known by us." A Zen poet around 700 A.D. expressed in beautiful poetic imagery the way man's observership complements the world:

The wild geese do not intend to cast
their reflection;
The water has no mind to receive
their image.[24]

Besides observership, something corresponding to the Anthropic principle is found in Aristotle and reaches a fuller, more explicit expression in the Middle

Ages. Aristotle, speaking of man's relation to nature, says, "We use everything as if it were there for our sake. We are also in a sense an end."[25] He adds "in a sense" because nature is ordered not only to man but also to the common end of the universe mentioned earlier. With the Anthropic principle, modern scientists show how assuming man to be the goal of nature accounts for certain properties of matter. Aristotle in a similar vein, argues in his biological works that we can best understand the human body in light of its relation to the human mind.[26]

By the Middle Ages the equivalent of the Anthropic principle becomes even clearer. Aquinas writes, "Those creatures that are less noble than man exist for the sake of man," serving either his bodily needs or his intellectual needs since by studying them he can eventually ascend to a knowledge of God.[27] Man is the center of the universe in the pre-scientific world view because he has a mind, *not* because the planets and stars were thought to revolve around the earth.

Aristotle sees in the world ample evidence for God. Because he sees purpose in nature, he concludes that there is a mind behind it: "Nature is divinely planned, though not itself divine." Further, Aristotle acknowledges that nature is full of beauty, a beauty that cannot be explained by material necessity or by chance. He therefore refers the beauty of nature to the Divine Artist. Thus, while advising neophyte biologists not to recoil in aversion from examining the humbler animals, he notes, "For if some have no graces to charm the sense, yet even these, by disclosing to intellectual perception the artistic spirit that designed them, give immense pleasure to all who can trace links of causation, and are inclined to philosophy." The analogy between God and the artist is, in fact, as old as Western philosophy itself. For Thales of Miletus, the first of the Greek philosophers, declares: "Of all things that are, the most ancient is God, for He is uncreated. The most beautiful is the universe, for it is God's workmanship."[28]

As for method, Aristotle acknowledges the primacy of common experience, just like the scientists of the New Story. Unlike Galileo and Bacon, he discards none of the certitudes of common experience. On the contrary, he grounds his whole investigation of nature upon them. Aristotle also sees wonder, not utility, as the catalyst for science:

"It is owing to their wonder that men both now begin and at first began to philosophize; they wondered originally at the obvious difficulties, then advanced little by little. . . . Since they philosophized in order to escape ignorance, evidently they were pursuing science in order to know, and not for any utilitarian end."[29]

And finally, Aristotle's attitude toward his predecessors can be seen in his treatise *On the Soul* where he says, "For our study of soul it is necessary . . . to call into council the views of those of our predecessors who have developed any opinion on the subject, in order that we may profit by whatever is sound in their suggestions and avoid their errors."[30]

He begins virtually every inquiry with a careful consideration of what others have said. Before he wrote his *Politics,* he analyzed the constitutions of 158 governments. And even *after* arguing to a major conclusion of his own, he often takes the trouble to consult again the views of others. For example, after discussing happiness, he writes, "We must consider it, however, in the light not only of our conclusions and our premises, but also of what is commonly said about it; for with a true view all the data harmonize, but with a false one the facts soon clash."[31] This also shows his respect for common experience.

With this outline we see that the New Story of science and the Aristotelian world view are surprisingly similar. Both assert that the mind is non-material; both recognize free choice in man, the superiority of spiritual goods, the beauty of nature, purpose in the universe, observership, the Anthropic principle, God, the primacy of common experience, wonder as the motive for science, and the wisdom of consulting one's predecessors. All of these points, denied by the Old Story of science, are points of unity between the New Story of science and the pre-scientific world view. Considering the great affinity between these two world views, it is not surprising to see Heisenberg explicitly refer to the concept of *"potentia* in the sense of Aristotelian philosophy" to explain the kind of being found in sub-atomic particles.[32] Heisenberg looks to Aristotelian physics rather than to Old Story materialism because the basic philosophy of the former is much closer to twentieth century science than is the latter.

But how were all these discoveries made so many centuries ago without the help of relativity, quantum physics, modern neuroscience, and astrophysics? Aristotle and the other thinkers of the pre-scientific tradition reached these truths on the basis of common experience. The basic principles of the New Story, attained through the specialized experience of modern science, are rediscoveries of long-known truths grounded in the common experience of all mankind. In this context Erwin Schrödinger writes, "Physical science in its present form . . . is the direct offspring, the uninterrupted continuation of ancient science."[33] From such a perspective, the Old Story of science appears as a 300-year detour from the mainstream of Western thought.

The elements of the New Story, then, bear the authority not only of the specialized experience of modern science but also the much greater authority of common experience. In this spirit physicist Eugene Wigner writes in his book *Symmetries and Reflections,* "The principal argument against materialism is not that it is incompatible with quantum theory. The principal argument is that thought processes and consciousness . . . cannot be denied."[34] The specialized experience of modern science has reminded contemporary man of things he should have not forgotten in the first place.

To be sure, there were mistakes in subsidiary conclusions of the Aristotelian world view. For example, Aristotle thought the sun revolved around the earth, that the matter composing the planets and stars was essentially different from all the earthly elements, and that in special cases life could generate itself spontaneously out of non-living matter. But to correct these errors does not require the rejection of any of the fundamental principles found both in Aristotle and in the New Story. No dis-

covery by Copernicus, or Galileo, or anyone else required science to deny free
choice, purpose, beauty, or God.

The remarkable agreement between the
ancient world view and contemporary science should not, however, lead us to believe there is nothing at all new about the
New Story. Our scientific world is signifi-
cantly different from that of the ancients.
They saw the universe as essentially
static. But in the contemporary understanding of the universe everything has a
history—even matter itself. The Big Bang
and grand unified theories of physics point
to a historical origin of even the most fundamental laws of physics. The New Story
presents us with a grander, more intimate
unity than ever envisioned by the ancients. Aristotle would have been astounded to learn that each element in our
bodies was manufactured in the heart of a
star billions of years ago. The elaborate
experiments, the sophisticated instruments, the subtle mathematics of modern
physics allow us to penetrate nature to a
depth undreamed of by Plato and Aristotle.
As a corollary, scientific technology has
transformed our culture economically, so-
cially, and politically. Because of the
power of this technology, modern man also
has more of a say in his own physical wel-
fare than ever before. Finally, only modern
man has the experience of living through

the Old Story. Materialism never dominated the ancient world view. The ancients never tasted the anguish of Camus, the dehumanization of the Old Story psychologies, or the forlornness of a universe thought to be godless, purposeless, and without beauty. Having lived with the Old Story for 300 years, modern man is in a better position to judge it.

The New Story of science rejoins modern man to the tradition of ancient wisdom. The same continuity holds in the arts. No great artist ever blasphemed his predecessors. Mozart admonishes, "People make a mistake who think that my art has come so easily to me. Nobody has devoted so much time and thought to composition as I. There is not a famous master whose music I have not studied over and over."[35]

Tradition is the ballast of a civilization. Without it we are tossed about by the arbitrary winds of fashion. Painter Jean Auguste Ingres challenges the cult of novelty that can be so destructive to the arts:

"Let me hear no more of that absurd maxim: 'We need the new, we need to follow our century, everything changes, everything is changed.' Sophistry—all of that! Does nature change, do the light and air change, have the passions of the human heart changed since the time of Homer? To claim that we can get along without study of the antique and the classics is either madness or laziness . . . It is the doctrine of those who want to produce without having worked, who want to know without having learned."[36]

Igor Stravinsky argues that, in music, tradition does not restrict the artist but furthers creativity and guarantees the continuity of the art:

"A real tradition is not the relic of a past that is irretrievably gone; it is a living force that animates and informs the present. . . . Far from implying the repetition of what has been, tradition presupposes the reality of what endures. It appears as an heirloom, a heritage that one receives on condition of making it bear fruit before passing it on to one's descendants. . . . A tradition is carried forward in order to produce something new. Tradition thus assures the continuity of creation."[37]

Tradition does not imply slavish copying of what has gone before. Agreeing with Stravinsky that tradition cannot be received passively, T.S. Eliot explains how it enables the artist to see the timeless in the temporal:

"If the only form of tradition, of handing down, consisted in following the ways of

the immediate generation before us in a
blind or timid adherence to its successes,
'tradition' should positively be discour-
aged. We have seen many such simple currents soon lost in the sand, and novelty is
better than repetition. Tradition is a matter of much wider significance. It cannot be
inherited, and if you want it you must ob-
tain it by great labour. It involves, in the
first place, the historical sense, which we
may call nearly indispensable to anyone
who would continue to be a poet beyond his
twenty-fifth year; and the historical sense
involves a perception, not only of the past-
ness of the past, but of its presence; the
historical sense compels a man to write not
merely with his own generation in his
bones, but with a feeling that the whole of
the literature of Europe from Homer and
within it the whole of the literature of his
own country has a simultaneous existence
and composes a simultaneous order. This
historical sense which is a sense of the
timeless and of the temporal together, is
what makes a writer traditional. And it is
at the same time what makes a writer
most acutely conscious of his place in time,
of his own contemporaneity."[38]

Tradition, as the word's etymology indicates, is a deliberate "handing over" of
something timeless in its truth or universal in its beauty. He is the most worthy

heir who can make the legacy bear new fruit. Both the artist struggling for beauty and the scientist striving for truth must labor with the help of their contemporaries and their predecessors. Beauty and truth are common goods and demand a common effort.

Tradition joins all mankind in a common good that spans the centuries. Without it there would be no Stravinskys, no Einsteins, no Mozarts, no Eliots—an overwhelming loss. For their symphonies, their theories, their poems are gifts to all mankind. The men who have gone before us augment our experience and multiply our faculties. We can see with their eyes, we can understand with their minds, and we can feel with their hearts. In this way only can we become most fully what men can be.

VIII

THE PRESENT

So far we have contrasted the Old Story of science with the New Story of science on the topics of matter, mind, beauty, God, psychology, the world, and the past. Now we are ready to compare these two Stories with each other as wholes. We expect a world view to provide three things: vastness, unity, and light. Its vastness ought to preserve the richness of all we experience; its unity ought to pull things together in a simple way; and its light ought to make sense out of things otherwise obscure.

In each of these categories the Old Story falls short of the New Story. As for vastness, the Old Story is narrow. According to it, nothing can be known except matter and its properties. It has difficulty accommodating moral, aesthetic, and intellectual values, purpose, and God. Erwin

Schrödinger, in *Mind and Matter,* complains about the impoverished picture given by the "world of science." If we consider his words carefully we realize he is speaking about the Old Story:

"The 'world of science' has become so horribly objective as to leave no room for the mind and its immediate sensations. . . . The world of science lacks or is deprived of everything that has a meaning only in relation to the consciously contemplating, perceiving and feeling subject. I mean in the first place the ethical and aesthetical values, any values of any kind, everything related to the meaning and scope of the whole display. . . . No personal god can form part of a world-model that has become accessible only at the cost of removing everything personal from it."[1]

Roger Sperry concurs on the narrowness of Old Story materialism:

"Consciousness, free will, and values: three long-standing thorns in the hide of science. Materialist science couldn't cope with any of them, even in principle. It's not just that they're difficult. They're in direct conflict with the basic models. Science has had to renounce them—to deny their existence or to say that they're beyond the domain of science. For most of us, of course, all three are among the most important things in life. When science proceeds to deny their importance, even their existence, or to say that they're beyond its domain, one has to wonder about science."[2]

In *Nature and the Greeks,* Schrödinger elaborates on the limitations of the Old Story science:

"The scientific picture of the real world around me is very deficient. It gives a lot of factual information, puts all our experience in a magnificently consistent order, but it is ghastly silent about all and sundry that is really near to our heart, that really matters to us. It cannot tell us a word about red and blue, bitter and sweet, physical pain and physical delight; it knows nothing beautiful or ugly, good or bad, God and eternity. Science sometimes pretends to answer questions in these domains, but the answers are very often so silly that we are not inclined to take them seriously."[3]

Werner Heisenberg agrees that Old Story materialism is narrow:

"The nineteenth century developed an extremely rigid frame for natural science which formed not only science but also the general outlook of great masses of people. . . . Matter was the primary reality.

The progress of science was pictured as a crusade of conquest into the material world. Utility was the watchword of the time. . . .

This frame was so narrow and rigid that it was difficult to find a place in it for many concepts of our language that had always belonged to its very substance, for instance, the concepts of mind, of the human soul or of life. . . . It was especially difficult to find in this framework room for those parts of reality that had been the object of the traditional religion and seemed now more or less imaginary. . . . Hostility of science toward religion developed . . . only the ethical values of the Christian religion were excepted from this trend, at least for the time being. Confidence in the scientific method and in rational thinking replaced all other safeguards of the human mind."[4]

Because of its rigid framework, the Old Story never raises certain questions; many issues become foregone conclusions. Freud, for example, assumes that God is a fiction and then tries to explain why people believe in this fiction. Arguments for or against God are by-passed. Darwin follows a similar procedure. In *The Descent of Man,* he does not argue whether there is a God or not. He speaks only of the *belief* in God and immortality, and then speculates as to how such beliefs might have arisen out of "long-continued culture."[5] These are natural and logical procedures if we assume that only matter is real.

The New Story, by contrast, includes the truths of the material world and opens the way to investigate spiritual realities as well. Mind is not an embarrassment in the New Story. On the contrary, the autonomy of mind and its centrality in the universe are the theme. The New Story in no way diminishes the importance or validity of brain research, molecular biology, or studies of animal behavior. It simply tells us not to expect those disciplines to explain away the mind. Heisenberg comments: "Those aspects of reality characterized by the words 'consciousness' and 'spirit' can be related in a new way to the scientific conception of our time."[6]

As for unity, the Old Story promises an appealing simplicity: everything is matter. But this turns out to be an illusory oversimplification. The Old Story never succeeded in pulling the sciences together. Heisenberg comments on the materialism and mechanism that characterized the science of the nineteenth century:

"Mechanics was the methodological example for all science. Although this view of nature had decisively advanced the development of science, it was soon seen that it was incapable of creating a durable unity of its different branches. . . . Finally, no suitable place could be found in this view of nature for that great realm of reality comprising mental processes."[7]

Besides its failure to unify the sciences, the Old Story creates a divorce between the sciences and the arts. The sciences it considers the domain of truth, but truth devoid of all values. The arts it considers the realm of private values with no foundation in truth (see Chapter III). The Old Story also produces antagonism between science and religion (see Chapter IV); it sets the specialized experience of science at odds with common experience (see Chapter VI); and the Old Story sees no continuity between modern science and ancient thought (see Chapter VII). Finally, the materialism of the Old Story ultimately undermines science itself (see Chapter VI). In this sense the Old Story does not even agree with itself. A greater lack of unity is difficult to imagine.

The New Story, on the other hand, brings a startling new unity to the sciences, a unity that is at once profound and far-reaching. In the twentieth century, physics, neuroscience, and humanistic psychology all converge on the same principle: mind is not reducible to matter. And the primacy of mind connects relativity with quantum mechanics, brain research with the Big Bang, the strength of nuclear forces with the size of the universe. Besides unifying the sciences, the New Story also re-unites the sciences with the arts since each studies and pursues beauty by different paths, as seen in Chapter III. The New Story also allows for conciliation of science and religion (see Chapter IV) and shows how common experience is the foundation of all knowledge, including science itself (Chapter VI). Finally, the New Story rejoins modern science with the tradition of ancient wisdom. This was seen in the quotations from pre-scientific thinkers (Chapter VII) who, drawing on the insights of common experience, understood the major principles of what is to us the New Story. The unity of the New Story is remarkable.

As for light, there are many things the Old Story has no hope of explaining. Beauty, for example. Materialism allows only two kinds of cause—chance and necessity. But as we saw in Chapter IV, neither one can account for the beauty found in nature. This is why Darwin finds that the Old Story offers no light on the question: "How the sense of beauty first developed in the mind of man . . . is a very obscure subject."[8]

As we saw in Chapters I and IV, materialism cannot even provide a complete understanding of matter. And as for non-material realities, the Old Story offers no light. It cannot understand the mind because it conceives of it as a by-product of matter. The reduction of mind to matter has taken many forms under the program of the Old Story. Some assume that matter must have evolved into mind. Others hope that neuroscience will eventually account for mind in terms of physics and chemistry. Still others adopt a materialistic model of how the mind operates. The proponents of "artificial intelligence" look for rules of thought that will enable them to produce genuine understanding and reasoning in a piece of hardware. To speak of "laws" of the mind or of the "structure" of the mind, however, is already to assume a material model.

John Eccles declares:

"I am appalled by the naiveté of the statements and arguments that are made by the proponents of the computer simulation of man. . . . There is no evidence whatsoever for the statements made that, at an adequate level of complexity, computers also would achieve self-consciousness."

Joseph Weizenbaum, Professor of Computer Science at Massachusetts Institute of Technology, also protests that "An entirely too simplistic notion of intelligence has dominated both popular and scientific thought. . . . Man is not a machine. . . . Computers and men are not species of the same genus."[9]

When the Old Story is confronted by the impossibility of explaining spiritual realities by matter alone, it often takes refuge in the future, assuming the answer will come after another century or two of research. Eccles calls this "promissory materialism . . . extravagant and unfulfillable."[10] The New Story, on the other hand, is founded on what is known now, not on some imagined, future discovery. The Old Story is always a dream, a promise, a never-fulfilled hope. With the discovery of relativity and quantum physics, the materialistic program is seen to be unrealizable *in principle*. The vain expectation that matter will some day account for mind, or free choice, or beauty, is like the alchemist's dream of producing gold from lead.

Furthermore, materialism draws much of its light from the imagination. If only matter exists, then everything real is picturable. Part of the appeal of the Old Story is that it portrays the world in easily imagined terms: Euclidean space and Newton's tiny, billiard-ball atoms, for example. But new facts, assimilated by relativity and quantum physics, forced science to transcend picture thinking. As known by the twentieth-century physicist, the atom can be understood but it cannot be literally pictured. Likewise, the four-dimensional space of relativity theory can be understood but not pictured. The New Story is more intellectual, and therefore, more demanding than the Old Story.

An insistence on picture thinking would gravely diminish the light of science. Under such a restriction, many things could never be understood. Nineteenth-century physicist William Thompson (Lord Kelvin) runs into difficulty because he equates explanation with a *material, mechanical, imaginable* explanation:

"I never satisfy myself until I can make a mechanical model of a thing. If I can make a mechanical model, I understand it. As long as I cannot make a mechanical model all the way through I cannot understand, and that is why I cannot get the electro-magnetic theory of light."[11]

Admitting only materialistic, mechanical explanations would hobble science. Fortunately, the New Story has liberated science from the tyranny of the imagination and has restored the intellect to its rightful supremacy. The New Story can understand atoms, four-dimensional space, and non-material mind without limiting itself to materialistic, mechanistic, picturable models.

Materialism's lack of light has restricted modern science. The tendency is to force the subject to fit the method rather than adjusting the method to the subject matter. Materialism dismisses as illusion the greater part of common experience, including beauty, values, purpose, mind, and free choice. As Schrödinger said earlier, the Old Story cannot explain the most profound human concerns; it can only explain them away. Because of its lack of light, the materialistic, mechanistic program of the nineteenth century was destined for eventual failure. Einstein notes, "Science did not succeed in carrying out the mechanical program convincingly, and today no physicist believes in the possibility of its fulfillment."[12]

But if materialism is so deficient in vastness, unity, and light, then how did such a world view appear plausible, even as a methodology, to Galileo and other thinkers? The answer is that they were reacting against the scholastic science of their day. The decadent scholastics came about 300 years after the great theologians of the Middle Ages such as Aquinas and Bonaventure and were poorly-enlightened, derivative thinkers. Their sterile doctrines dominated the schools and academies of the late Renaissance. Bacon, Descartes, Hobbes, Locke, and many others who attended these schools complain of the scholastics' endless wrangling and hair-splitting disputes in which all is abstract and nothing is ever solved. Their teaching and writings often degenerated into polemic. Extending improperly the method of theology to philosophy and natural science, the scholastics invested the works of Aristotle with an exaggerated authority, almost tantamount to that of sacred scripture.

It was clear to any mind of integrity that a new model for science had to be formulated. In contrast to decadent scholasticism, where the metaphysical and supernatural seemed to absorb all else, the new model for science would have to turn back to the concrete, back to the indisputably real, back to the physical world of matter. The intellectual shift in world view from the Middle Ages to the Renaissance can be characterized very generally as a turning from the spiritual to the physical. At first the change is more of a new emphasis on matter rather than a rejection of the spiritual. This shift is reflected in the science, in the philosophy, and in the art of the Renaissance as contrasted with the Middle Ages. We have already seen in Chapter VI how Galileo assumes a methodological materialism. This notion is found in Francis Bacon also. Bacon considers natural science to be the "mother of the sciences" and the root of them all. And he declares that "in nature nothing really exists besides individual bodies. performing pure individual acts according to fixed law." He adds, "Matter rather than forms should be the object of our attention."[13] Only later does materialism become a dogma on its own, giving rise to a desire to explain all things by matter alone.

Next, to correct the scholastics' abuse of authority, the new model of science had to be founded neither upon revelation nor upon the authority of men. Natural science could be based only on the authority of experiment and of reason. Bacon justifiably complains that the decadent scholastics sterilized science with arbitrary authority:

"Philosophy and the intellectual sciences . . . stand like statues, worshipped and celebrated, but not moved or advanced. . . . For when men have once made over their judgments to others' keeping . . . from that time they make no enlargement of the sciences themselves, but fall to the servile office of embellishing certain individual authors and increasing their retinue. . . . Those very authors . . . have usurped a kind of dictatorship in the sciences."[14]

Galileo and others understandably opposed an obstinate and undiscerning attachment to Aristotle's physics. In a letter to Kepler, Galileo cites a famous example:

"Oh, my dear Kepler, how I wish that we could have one hearty laugh together! Here at Padua is the principal professor of philosophy, whom I have repeatedly and urgently requested to look at the moon and planets through my glass, which he pertinaciously refuses to do. Why are you not here? What shouts of laughter we should have at this glorious folly! And to hear the professor of philosophy at Pisa labouring before the Grand Duke with logical arguments, as if with magical incantations, to charm the new planets out of the sky."[15]

The professor thought he did not have to look through the telescope since the moons of Jupiter were not mentioned by Aristotle and therefore could not exist. Galileo says elsewhere that Aristotle himself, unlike his unworthy disciples, would not have distained the new evidence from sense experience.[16]

Descartes, in a letter to Mersenne in 1629, complains of the confusion of theology and natural science: "Theology has been so subjected to Aristotle that it is almost impossible to explain another philosophy without it seeming at first to be contrary to Faith." And physicist-mathematician Blaise Pascal also deplores "the blindness of those who advance authority alone as proof in physics instead of reason or experiment."[17]

Finally, in reaction to a neglect of man's natural needs, the new program for science had to be practical, not speculative only. "Human knowledge and power meet in one," writes Bacon.[18] And he criticizes ancient philosophers as having what is "characteristic of boys; they are prompt to prattle, but cannot generate; for their wisdom abounds in words but is barren of works."[19] Descartes also sets a practical goal for the new science:

"In place of the speculative philosophy taught in the Schools we can have a practical philosophy, by means of which, knowing the force and the actions of fire, water, air, of the stars, of the heavens, and of all the bodies that surround us—knowing them as distinctly as we know the various crafts of the artisans—we may in the same fashion employ them in all the uses for which they are suited, thus rendering ourselves the masters and possessors of nature."[20]

Methodological materialism, then, seemed plausible because it led to many new discoveries in natural science and suggested a cure for the intellectual ills of the day; it offered a way out of barren scholastic science. But it is always hazardous to form a view in reaction to another position. For in trying to compensate for a previous neglect of matter, Galileo and Descartes unwittingly initiated an eclipse of mind.

The Old Story of science, then, introduced some needed reforms, but its materialism set science on a wrong track. And it is very difficult to change a world view once it is established. An entrenched error in any part of science is always difficult to eradicate. Leopold Infeld writes, regarding the ether hypothesis of Newtonian physics, which, despite contradictions, some scientists persistently clung to: "Deep-rooted prejudices die hard. The physicist of the nineteenth century was not prepared to sacrifice the ether concept. He could not deny the evidence of experiment, but he could change his argument."[21] Unexamined pre-conceptions are a constant danger in science as Antoine Lavoisier, the father of modern chemistry, warned: "These suppositions, handed down from one age to another, acquire additional weight from the authorities by which they are supported, till at last they are received, even by men of genius, as fundamental truths."[22] Old Story materialism is just such a supposition.

But a mistake in world view is much more grave than a false hypothesis about ether or phlogiston. An error in a particular conclusion within one science is a minor evil, easily corrected by applying more basic principles. Much worse is an error in the very method of a science, for it will be multiplied every time the method is used and is much more difficult to cure. But the worst of all is an error in world view because it influences the methods of all the sciences as well as the attitudes and expectations in the arts, politics, religion, and every other phase of culture. A mistake in world view is necessarily an architectonic error, producing scores of other errors.

In *American Dilemma,* Swedish economist Gunnar Myrdal, after deploring pseudo-scientific racism, observes that:

"There must be still other countless errors of the same sort that no living man can yet detect, because of the fog within which our type of Western culture envelops us. Cultural influences have set up the assumptions about the mind, the body, and the universe with which we begin; pose the questions we ask; influence the facts we seek; determine the interpretation we give these facts; and direct our reaction to these interpretations and conclusions."[23]

Because it underlies so many aspects of our culture, Old Story materialism has taken on an aura of irrefutability. To assail it with arguments is futile: it is too comprehensive and too far-reaching to be vulnerable to that kind of criticism. The only way to overthrow it is to tell a whole New Story, an impossible task for an individual person. Yuri Glazov, a Soviet emigré, explains how the power of Marxism, for example, is not based on truth but on its magnitude and seeming monopoly on wholeness:

"The Soviet Union is the country of the official Marxist-Leninist ideology, which penetrates all levels of society, perhaps with greater thoroughness and stubbornness than religion in any theocracy. . . . Many people feel that they have been permeated by the Marxist indoctrination, but even this feeling of being overfilled with the Marxist ideology does not give them any strength to withstand it. For the Soviet person it is highly characteristic to criticize the system or its ideology on some details, but never in total. The more a person studies Marxism-Leninism, the more helpless he feels himself, with his knowledge and his background, to overthrow the Marxist heritage. He is particularly afraid to be left without any theory or ideology at all; he has nothing with which to replace it—no other decent and verified theory or world outlook."[24]

Marxism is only one species of Old Story materialism, but it illustrates well the difficulty of escaping any entrenched world view. A man can disagree with the world view prevailing in his culture. He can even fight against it; but he will never be a match for it as a social force.

Since the Renaissance, Old Story materialism has dominated our culture, influencing our choice of methods, our categories of thought, our imagery, and even our vocabulary. For example, since the time of Hobbes, much of psychology uses a mechanist vocabulary often without even realizing its origin in the machine model of man. Since Darwin, "survival" is evoked by practically everyone to account for practically everything.

A world view is enormously powerful and all-pervasive. But it is not the most fundamental thing in human knowledge. The common experience of all men is the absolute foundation of human knowledge. Why, then, was no appeal made to common experience in order to correct the narrowness of materialism? This was unthinkable because the Old Story undermined and discredited the authority of common experience, as we saw in Chapter VI. Thus, once established, the Old Story began to re-interpret the data of common experience, emphasizing certain parts of it and dismissing others. The Old Story began to tell scientist and layman alike what is real and what is not, where to look for answers, what to expect, and how to interpret the results. Old Story materialism in this way gradually usurped the office of common experience itself.

Because of its rejection of common experience, the Old Story could be overcome by one authority alone: specialized scientific experience, the source of the New Story. The New Story of science, while affirming the autonomy and primacy of mind, is nowhere founded on a plea that such a view offers more "consolation" than materialism. If the Old Story were true, we would have to learn to live with it, no matter how nihilistic or discomforting. The Old Story has been found wanting because the facts of relativity, quantum physics, neuroscience, and astrophysics cannot be fit into such a scheme. Eugene Wigner insists that in twentieth-century physics "the very study of the external world led to the conclusion that the content of the consciousness is an ultimate reality."[25] The Old Story must be replaced, then, not because it offers us no consolation, but because it is false. Materialism, having received the fullest possible trial in all branches of science, has simply failed the test of experience.

Right now our culture is between the Old Story and the New Story. As we have seen in previous chapters, the physics and the neuroscience of the twentieth century acknowledge the autonomy of the mind and its irreducibility to matter. Other disciplines still remain largely under the influence of the Old Story. Some scientists are clear and consistent champions of the New Story. They were the first to go beyond the Old Story because their major scientific discoveries led them out of materialism. Other thinkers today continue to work almost exclusively within the Old Story perspective. Still others acknowledge some elements of the New Story but try to retain the general framework of the old world view. But the internal coherence of the Old Story and of the New Story make compromise without contradiction highly unlikely.

As for the future, many things yet remain to be discovered about matter, but materialism itself seems to be played out. It can subsist but probably only at the expense of becoming progressively narrower, more dogmatic, and more obscurantist. The New Story of science, on the other hand, holds great promise for the future. We have seen that the New Story has already transformed modern physics. It is to be expected that the disciplines that have patterned themselves after the old physics will eventually adjust themselves to the *new* physics and all of its implications. The primacy of mind gives all the sciences a fresh perspective and a new light. The New Story promises to liberate and illuminate every discipline, making possible a genuine rebirth in our age.

We can expect contemporary philosophy, encouraged by the New Story, to turn from its intellectual despair to a healthy and vigorous search for wisdom based on the certitudes of common experience. As regards religion, the future of the New Story seems to imply a return to theism in our culture and the reaffirmation of the spiritual side of man's nature. Finally, concerning the arts, the New Story removes from psychology and cosmology the causes of alienation and meaninglessness, replacing them with purpose, God, beauty, the spiritual goods, and the dignity of man. Thus, the arts once more have something to celebrate and the poet something worthy of his song: Man is at home in the universe.

NOTES

INTRODUCTION

1. Thomas Berry, *The New Story* (Chambersburg, Pa,; Anima Publications, 1978), p. 1. We are indebted to Berry for the metaphors old story and new story, which we have developed in our own way, however.

Also, we have taken the notion "world view of science" not from Thomas Kuhn's *The Structure of Scientific Revolutions,* but from physicists such as Heisenberg, Weizsacker, and Schrödinger. We do not hold that scientific theories are sociologically determined or that science is ultimately reducible to conventionalism.

2. Bertrand Russell, "A Free Man's Worship," in *Why I Am Not A Christian* (New York: Simon & Schuster, 1957), p. 107.

3. Henry Margenau, *The Miracle of Existence* (Woodbridge, Connecticut: Ox Bow Press, 1984), p. 22. Werner Heisenberg, *Physics and Philosophy* (New York: Harper & Row, 1958), p. 59.

4. Harold J. Morowitz, *The Wine of Life and Other Essays on Societies, Energy and Living Things* (New York: St. Martin's Press, 1979), p. 33.

I MATTER

1. Isaac Newton, *Opticks* (New York: Dover, 1952), p. 400. Isaac Newton, *Principia,* trans. Florian Cajori (Berkeley & Los Angeles: University of California Press, 1934), p. 399.

2. Newton, *Principia,* p. 6.

3. Newton, *Opticks,* p. 400.

4. Newton, *Principia,* Preface to the First Edition, pp. xvii–xviii.

5. Max Born, *Physics in My Generation* (London & New York: Pergamon Press, 1956), p. 48.

6. John A. Wheeler, "Genesis and Observership," in *Foundational Problems in the Special Sciences,* ed. Robert E. Butts and Jaakko Hintikka (Dordrecht, Holland: Reidel, 1977), pp. 5–6.

7. Eugene Wigner, *Symmetries and Reflections* (Bloomington: Indiana University Press, 1967), p. 189.

8. P. 171.

II MIND

1. Thomas H. Huxley, *Selections From The Essays of T.H. Huxley,* ed. Alburey Castell (New York: Appleton-Century-Crofts, 1948), p. 19.

2. William Kingdon Clifford, "Mind and Body," *Lectures and Essays,* ed. Leslie Stephen and F. Pollack (London: Macmillan, 1879), II, p. 56.

3. Thomas H. Huxley, quoted by Gordon Rattray Taylor, *The Natural History of Mind* (New York: Dutton, 1979), p. 4.

4. Huxley, *Selections,* p. 21.

5. Charles Sherrington, *Man on His Nature* (Cambridge: Cambridge University Press, 1975), p. 230.

6. John Eccles, *Facing Reality* (Berlin: Springer-Verlag, 1970), p. 162.

7. Pp. 1–2.

8. P. 55.

9. Sherrington, p. 238.

10. Quoted in Wilder Penfield, *The Mystery of the Mind: A Critical Study of Consciousness and the Human Brain* (Princeton: Princeton University Press, 1975), p. 4.

11. William Kaufmann, *The Cosmic Frontiers of General Relativity* (Boston: Little, Brown, 1977), p. 76.

12. Penfield, *The Mystery of the Mind* (Princeton: Princeton University Press, 1975), p. 22.

13. P. 51.

14. P. 52.

15. P. 53.

16. P. 76.

17. Pp. 76–77.

18. P. 55.

19. Pp. 78, 77.

20. Eccles, *Facing Reality*, p. 4.

21. P. 120.

22. Carl F. von Weizsacker, *The World View of Physics* (Chicago: University of Chicago Press, 1949), p. 203.

23. Roger Sperry, "Interview," *Omni,* August 1983, p. 72. Sperry rejects the attempt to reduce man's mental experience to an epiphenomenon of brain physiology. He asserts strongly that mental events are real and causal. We do not mean to imply that he espouses an unqualified freedom of the will or individual immortality.

24. P. 72.

25. Penfield, p. 49.

26. John Eccles, *The Human Mystery* (New York: Springer-Verlag, 1979), p. 227.

27. Penfield, pp. 80, 75–76.

28. P. 79.

29. Adolf Portmann, *New Pathways in Biology* (New York: Harper & Row, 1964), p. 27.

30. Penfield, pp. 80, 62.

31. Pp. xiii, 85.

32. Eccles, *Facing Reality*, p. 174.

III   BEAUTY

1.   Louis de Brogile, *Savants et Découvertes* (Paris: Albin Michel, 1951), pp. 378–79, our translation. Descartes, from a letter to Mersenne of 18 March 1630, quoted by Tartarkiewicz, *History of Aesthetics,* ed. D. Petsch (The Hague and Paris: Mouton, 1974), III, p. 373. Baruch Spinoza, letter to H. Boxel, Sept. 1674, quoted by Tartarkiewicz, *History of Aesthetics,* III, p. 380.

2.   Charles Darwin, *The Origin of Species and The Descent of Man* (New York: Modern Library, n.d.), p. 147. Sigmund Freud, *Civilization and Its Discontents,* trans. James Strachey (New York: Norton, 1962), p. 30.

3.   James Watson, *The Double Helix* (New York: Mentor, 1968), pp. 131, 134.

4.   Richard Feynman, *The Character of Physical Law* (Cambridge: M.I.T. Press, 1965), p. 171. Werner Heisenberg, "The Meaning of Beauty in the Exact Sciences," in *Across the Frontier* (New York: Harper & Row, 1974), p. 183.

5.   Heisenberg, p. 175. Erwin Schrödinger, *Nature and the Greeks* (Cambridge: Cambridge University Press, 1954), p. 23. Albert Einstein, quoted by S. Chandrasekhar, "Beauty and the Quest for Beauty in Science," *Physics Today* 32 (July, 1979), p. 26.

6. P.A.M. Dirac, "The Evolution of the Physicist's Picture of Nature," *Scientific American* 208 (May 1963), p. 47. Sir George Thomson, *The Inspiration of Science* (Oxford: Oxford University Press, 1961), p. 17.

7. Richard Feynman and Murray Gell-Mann, *Physical Review* 109 (1958), p. 193. Murray Gell-Mann, quoted by Horace F. Judson, *Search for Solutions* (New York: Holt, Rinehart & Winston, 1980), p. 22.

8. S. Chandrasekhar, p. 27.

9. Albert Einstein, "Autobiographical Notes," in *Albert Einstein: Philosopher-Scientist,* ed. Paul Schilpp (New York: Harper & Row, 1959), p. 33.

10. Roger Penrose, "Black Holes," in *The State of the Universe,* ed. Geoffrey Bath (Oxford: Clarendon Press, 1980), p. 128.

11. Henri Poincaré, *Science et Méthode* (Paris: Flammarion, 1949), p. 16, our translation. Heisenberg, "Meaning of Beauty," p. 167.

12. Albert Einstein and Leopold Infeld, *The Evolution of Physics* (New York: Simon & Schuster, 1938), p. 313. Heisenberg, "Meaning of Beauty," p. 167. John A. Wheeler, "The Universe as a Home for Man," *American Scientist* 62 (Nov.-Dec. 1974), p. 688. Heisenberg, *Physics and Philosophy,* p. 133. Newton, *Principia,* p. 13.

13. Thomson, p. 18. Chandrasekhar, p. 29.

14. Newton, "Rules of Reasoning," *Principia,* p. 399. Richard Feynman, p. 173. Wheeler, p. 688. Max Born, *The Restless Universe* (New York: Dover, 1951), p. 54. Werner Heisenberg, *Physics and Beyond* (New York: Harper & Row, 1972), pp. 68–69.

15. Chandrasekhar, p. 25. David Bohm, in *Towards a Theoretical Biology,* ed. C.H. Waddington (Chicago: Aldine, 1969), p. 50. Henri Poincaré, *The Value of Science* (New York: Dover, 1958), p. 8.

16. Von Weizsacker, *The World View of Physics,* p. 179.

17. C.P. Snow, "The Moral Un-neutrality of Science," in his *Public Affairs* (New York: Scribner's, 1971), p. 189.

18. Albrecht Dürer, in *Artists on Art,* ed. Robert Goldwater and Marco Treves (New York: Pantheon, 1958), p. 82. Vincent Van Gogh, in Walter Sorell, *The Duality of Vision* (Indianapolis: Bobbs-Merrill, 1970), p. 137. Johannes Brahms, in Joseph Machlis, *The Enjoyment of Music* (New York: Norton, 1963), p. 177. Henri Matisse, in *Artists on Art,* p. 411.

19. Albrecht Dürer, in Tartarkiewicz, *History of Aesthetics,* p. 257. Leon Battista Alberti, in *Artists on Art,* p. 36.

20. Christoph Gluck, *Composers on Music,* ed. Sam Morgenstern (New York: Bonanza Books, 1956), p. 66. Henri Matisse, in *Artists on Art,* pp. 411, 412.

21. Edouard Manet, quoted by Pierre Schneider, *The World of Manet* (New York: Time-Life Books, 1968), p. 104. Leonardo DaVinci, in *Artists on Art,* p. 53.

22. Aaron Copland, *What to Listen for in Music* (New York: McGraw-Hill, 1957), p. 78. Nicolai Rimsky-Korsakov, in *Composers on Music,* p. 275.

23. Steven Weinberg, "Science's 'Parallels with Art'—a Physicist's View," *U.S. News and World Report,* 8 Sept. 1980, p. 68.

24. P. 68.

25. Werner Heisenberg, *Philosophical Problems of Nuclear Science* (New York: Pantheon, 1952), p. 75.

IV GOD

1.   Francis Bacon, *The New Organon* (Indianapolis: Bobbs-Merrill, 1960), p. 121. René Descartes, *Médi-*

tations, in *Descartes Oeuvres et Lettres,* ed. André Bridoux (Paris: Editions Gallimard, 1953), p. 303, our translation.

2.   Isaac Newton, *Newton's Philosophy of Nature: Selections From His Writings,* ed. H.S. Thayer (New York: Macmillan, 1953), p. 47.

3.   Pierre-Simon Laplace, quoted by Eric Temple Bell, *Men of Mathematics* (New York: Simon & Schuster, 1937), p. 181. Carl F. Gauss, quoted in Bell, p. 240.

4.   Sigmund Freud, *Civilization and Its Discontents,* trans. James Strachey (New York: Norton, 1962), p. 28. Sigmund Freud, *The Future of an Illusion,* trans. W.D. Robson-Scott (Garden City: Doubleday, 1964), pp. 30, 81.

5.   Dennis W. Sciama, "The Origin of the Universe," in *The State of the Universe,* ed. Geoffrey Bath (Oxford: Clarendon Press, 1980), p. 3. Wheeler, "Genesis and Observership," p. 17.

6.   Steven Weinberg, *The First Three Minutes* (New York: Basic Books, 1977), p. 154.

7.   Sidney A. Bludman. "Thermodynamics and the end of a closed Universe," *Nature,* 308 (22 March 1984), p. 322. Wheeler "Genesis and Observership," p. 15.

8.   Joseph Silk, *The Big Bang: The Creation and Evolution of the Universe* (San Francisco: W.H. Freeman, 1980), p. 312. Robert Jastrow, *God and the Astronomers* (New York: Norton, 1978), p. 14.

9.   Edmund Whittaker, quoted in Jastrow, pp. 111–12. Edward Milne, quoted in Jastrow, p. 112.

10.   Freud, *Future of an Illusion,* p. 81.

11.   Brandon Carter, "Large Number Coincidences and the Anthropic Principle in Cosmology," in *Confrontation of Cosmological Theories with Observational Data,* ed. M.S. Longair (Dordrecht, Holland, and Boston: Reidel, 1974), p. 291.

12.   P. 291.

13. Steven W. Hawking, "The Anisotropy of the Universe at Large Times," in *Confrontation of Cosmological Theories,* pp. 285–86.

14. Wheeler, p. 21.

15. Carter, p. 294.

16. Wheeler, p. 18.

17. Freeman Dyson, *Disturbing the Universe* (New York: Harper & Row), p. 250.

18. P. 251. For an extensive discussion of these and other "lucky accidents," see the excellent review article by B.J. Carr and M.J. Rees, "The Anthropic Principle and the Structure of the Physical World," *Nature* 278 (12 April 1979), pp. 605–12.

19. Wheeler, p. 21.

20. Dyson, p. 250.

21. Wheeler, pp. 7, 21, 27. Erwin Schrödinger, *Mind and Matter* (Cambridge: Cambridge University Press, 1959), p. 2.

22. Heisenberg, *Physics and Philosophy,* p. 81.

23. W.A. Bentley, *Snow Crystals* (New York: Dover, 1962), p. i.

24. P. i.

25. Gary Taubes, "The Snowflake Enigma," *Discover* (Jan. 1984), pp. 75–78.

26. Adolf Portmann, *New Pathways in Biology,* p. 101.

27. P. 102.

28. Darwin, *The Origin of Species and The Descent of Man,* p. 878.

29. Bohm, in Waddington, p. 50. Adolf Portmann, *Animal Forms and Patterns* (New York: Schocken, 1967), p. 28.

30. Henry David Thoreau, *Journal,* in Thoreau, *In Wilderness is the Preservation of the World,* ed. Eliot Porter (New York: Ballantine Books, 1967), p. 70.

31. Henry Margenau, *The Miracle of Existence*

(Woodbridge, Connecticut: Ox Bow Press, 1984), pp. 29–30.

32. Henry David Thoreau, *Thoreau On Man and Nature,* ed. Arthur Volkman (Mount Vernon, NY: Peter Pauper Press, 1960), p. 6.

33. Thomas Mann, *Death in Venice and Seven Other Stories* (New York: Vintage, 1954), p. 72. Ralph Waldo Emerson, quoted in *The New Dictionary of Thoughts* (London & New York: Classic Publishing, 1936), p. 41.

34. Elizabeth Barrett Browning, "The Dead Pan," in *The Complete Works of Elizabeth Barrett Browning,* ed. Charlotte Porter and Helen A. Clarke (New York: Crowell, 1900), III, p. 157.

V MAN AND SOCIETY

1. Thomas Hobbes, *Leviathan* (Indianapolis: Bobbs-Merrill, 1958), p. 68.

2. P. 87.

3. P. 107.

4. P. 139.

5. Sigmund Freud, "Psychoanalysis and Telepathy," in *Standard Edition of the Complete Psychological Works of Sigmund Freud,* ed. James Strachey (London: Hogarth Press, 1955), 18, p. 179. Sigmund Freud, *The Future of an Illusion,* p. 57.

6. Sigmund Freud, *Civilization and Its Discontents,* p. 29.

7. P. 23.

8. P. 22.

9. Pp. 26–28, *passim.* Lawrence John Hatterer, "Work Identity: A New Psychotherapeutic Dimension," *Psychiatric Spectator,* 2, no. 7, (1965), p. 12.

10. Freud, *The Future of an Illusion,* pp. 5, 12, 3.

11. P. 62.

12. John B. Watson, *Behaviorism* (Chicago: University of Chicago Press, 1930), p. 5.

13. Pp. 5, 98, 242, 224.

14. Pp. 112–13.

15. P. 6.

16. John B. Watson, *Psychology from the Standpoint of a Behaviorist* (Philadelphia: Lippincott, 1919), pp. 1–2.

17. P. v.

18. Frank T. Severin, *Discovering Man in Psychology: A Humanistic Approach* (New York: McGraw-Hill, 1973), p. 5.

19. Irvin L. Child, *Humanistic Psychology and the Research Tradition: Their Several Virtues* (New York: John Wiley, 1973), p. 13.

20. P. 15.

21. Severin, pp. 5, 6.

22. Carl Rogers, "Toward a Science of the Person," in *Behaviorism and Phenomenology: Contrasting Bases For Modern Psychology,* ed. T.W. Wann (Chicago: University of Chicago Press, 1964), p. 119.

23. Viktor Frankl, *The Unconscious God* (New York: Simon & Schuster, 1975), p. 95.

24. P. 111.

25. Rollo May, *Psychology and the Human Dilemma* (New York: Norton, 1979), p. 182.

26. Rogers, p. 119. Severin, p. 6.

27. Sperry, "Interview," p. 71.

28. May, p. 208.

29. Freud, *Civilization and Its Discontents,* pp. 26–27.

30. Abraham H. Maslow, *Toward a Psychology of Being* (New York: Van Nostrand Reinhold, 1968), p. 200.

31. Viktor Frankl, *Psychotherapy and Existentialism* (New York: Simon & Schuster, 1967), p. 20.

32. Viktor Frankl, *Man's Search for Meaning* (New York: Pocket Books, 1959), pp. 157–58.

33. P. 213.

34. May, p. 176.

35. Frankl, *Man's Search for Meaning*, pp. 56–57.

36. Roger Sperry, "Mind, Brain and Humanist Values," in *New Views of the Nature of Man*, ed. John R. Platt (Chicago: University of Chicago Press, 1965), pp. 78, 82.

37. Maslow, p. 159.

38. P. 159.

39. Aleksandr Solzhenitsyn, *The Gulag Archipelago Two*, trans. Thomas P. Whitney (New York: Harper & Row, 1975), pp. 605–6.

40. Frankl, *Man's Search for Meaning*, p. 62.

41. P. 104.

42. Maslow, p. 159.

VI THE WORLD

1. Bacon, *The New Organon*, p. 21.

2. P. 22.

3. Pp. 22, 25.

4. Galileo, in Edwin Arthur Burtt, *The Metaphysical Foundations of Modern Physical Science* (Garden City: Doubleday, 1954), pp. 76, 75.

5. Galileo, *Discoveries and Opinions of Galileo*, trans. Stillman Drake (Garden City: Doubleday, 1957), p. 274.

6. P. 274.

7. P. 274. Galileo, in Burtt, p. 85. Galileo, pp. 276–77.

8. Descartes, *Philosophical Writings,* ed. Norman K. Smith (New York: Modern Library, 1958), pp. 51, 119.

9. Jean-Paul Sartre, *Existentialism and Human Emotions,* trans. Bernard Frechtman (New York: Citadel Press, 1957), p. 36. Albert Camus, *The Myth of Sisyphus,* trans. Justin O'Brien (New York: Vintage Books, 1955), p. 10.

10. George Berkeley, *A Treatise Concerning The Principles of Human Knowledge* (Indianapolis: Bobbs-Merrill, 1957), p. 30.

11. George Berkeley, "Philosophical Commentaries, No. 24," in *The Works of George Berkeley,* ed. A.A. Luce and T.E. Jessop (London and Edinburgh: Thomas Nelson, 1948), I, p. 10.

12. David Hume, *A Treatise of Human Nature* (Oxford: Clarendon Press, 1888), pp. 180–87.

13. Immanuel Kant, *Prolegomena to Any Future Metaphysics* (Indianapolis: Bobbs-Merrill, 1950), p. 67.

14. Bacon, p. 22.

15. Sartre, p. 36. Camus, p. 13.

16. Pp. 11, 38.

17. P. 14.

18. Friedrich Nietzsche, *Beyond Good and Evil,* trans. Marianne Cowan (Chicago: Henry Regnery, 1955), p. 120.

19. Franz Marc, in *Artists on Art,* p. 445.

20. Piet Mondrian, in *Artists on Art,* p. 428. Edward Wadsworth, in *Artists on Art,* p. 458.

21. Gustave Flaubert, in *Writers on Writing,* ed. Walter Allen (New York: Dutton, 1949), pp. 136–37.

22. Anton Chekov, in *Writers on Writing,* p. 137.

23. Anatole France, "La Vie Litteraire," in *Critical Theory Since Plato,* ed. Hazard Adams (New York: Harcourt Brace, 1971), p. 671.

24. Erwin Schrödinger, *Mind and Matter* (Cambridge: Cambridge University Press, 1958), p. 51.

25. Eccles, *Facing Reality,* p. 52.

26. Heisenberg, *Physics and Philosophy,* p. 149.

27. Einstein and Infeld, *The Evolution of Physics,* p. 277.

28. Heisenberg, *Physics and Philosophy,* p. 168. Niels Bohr, *Essays, 1958–1962, On Atomic Physics and Human Knowledge* (New York: Interscience, 1963), p. 1.

29. Heisenberg, *Physics and Philosophy,* p. 200.

30. P. 200.

31. Sperry, "Interview," p. 98.

32. Heisenberg, *Physics and Philosophy,* p. 200.

33. Schrödinger, *Nature and the Greeks* (Cambridge: Cambridge University Press, 1951), pp. 55–56.

34. Leopold Infeld, *Albert Einstein: His Work and Its Influence on Our World* (New York: Scribner's, 1950), p. 41.

35. P. 84.

36. Schrödinger, *Science and Humanism* (Cambridge: Cambridge University Press, 1951), p. 7. Morris Kline, *Mathematics: The Loss of Certainty* (New York: Oxford University Press, 1980), p. 284. Schrödinger, *Science and Humanism,* pp. 5, 9.

37. Carl von Weizsacker, *The World View of Physics,* p. 202. Frankl, *Doctor and the Soul,* p. 29.

38. Mozart, in *Composers on Music,* p. 83.

39. Roger Sessions, in *Composers on Music,* p. 502.

40. Goethe, in *Writers on Writing,* pp. 35–36.

VII   THE PAST

1.   Bacon, *The New Organon,* p. 46.

2.   Ernst Mach, "On Instruction in the Classics and the Sciences," in *Popular Scientific Lectures,* trans. Thomas J. McCormack (La Salle, Illinois: Open

Court, 1943), p. 344. Our citation is Schrödinger's translation, *Nature and the Greeks* (Cambridge: Cambridge University Press, 1954), p. 18.

3.   Descartes, *Philosophical Writings,* pp. 8, 105.

4.   Umberto Boccioni, in *Artists on Art,* pp. 434–35.

5.   Werner Heisenberg, "Changes in the Foundations of Exact Science," *Philosophical Problems of Nuclear Science,* trans. F.C. Hayes (New York: Pantheon, 1952), p. 19.

6.   Eccles, *Facing Reality,* p. 6.

7.   Rogers, p. 129.

8.   Einstein, "Time, Space, and Gravitation," in *Out of My Later Years* (New York: Philosophical Library, 1950), p. 58.

9.   Infeld, p. 26.

10.   Louis de Broglie, *La Physique Nouvelle et les Quanta* (Paris: Flammarion, 1937), p. 13, our translation.

11.   Aristotle, *On The Soul,* Bk. II, Ch. 12, p. 580. All citations from Aristotle are in *The Basic Works of Aristotle,* ed. Richard McKeon (New York: Random House, 1941).

12.   P. 590.

13.   P. 592.

14.   Aristotle, *Nicomachean Ethics,* p. 971.

15.   Aristotle, *Politics,* p. 1278.

16.   See for example, Aristotle's *Politics,* Bk. VII, Ch. 1 and his *Nicomachean Ethics,* Bk. IX, Ch. 8.

17.   See for example, *Nicomachean Ethics,* Bk. X, Chs. 7 and 8.

18.   Aristotle, *On the Parts of Animals,* p. 657. Aristotle, *Physics,* p. 226.

19.   Aquinas, *The Pocket Aquinas,* ed. Vernon J. Bourke (New York: Pocket Books, 1960), p. 263.

20.   Aristotle, *Physics,* p. 251. In addition to *Physics,* Bk. II, Ch. 8, Aristotle has an extended dis-

cussion of the use of purpose as an explanation in biology in his *On the Parts of Animals,* Bk. I, Ch. 1.

21. Aristotle, *On The Soul,* p. 597.

22. Aristotle, *Metaphysics,* pp. 885–86.

23. P. 689. Aristotle, *Nicomachean Ethics,* p. 981.

24. Cicero, *De Natura Deorum,* trans. H. Rackham (Cambridge: Harvard University Press, 1933), p. 159. Epictetus, "The Discourses," *The Works of Epictetus,* trans. Thomas Higginson (New York: Nelson, 1890), p. 24. Augustine, *The City of God,* trans. G. Walsh, *et al.,* (Garden City: Doubleday, 1950), pp. 237–38. From *Zenrin Kushu,* an anthology of over five thousand two-line poems, compiled by Toyo Eicho (1429–1504); see Alan W. Watts, *The Way of Zen* (New York: Random House, 1957), pp. 117 and 181.

25. Aristotle, *Physics,* p. 240.

26. For instance, Aristotle argues that having hands, the most universal of tools, is a consequence of rather than a cause of man's superior intelligence, nature always giving organs to such animals as can make the best use of them. See *On the Parts of Animals,* Bk. IV, Ch. 10, beginning at line 687a6.

27. Aquinas, *Summa Theologica,* English trans. (New York: Benziger Brothers, 1947), I, p. 326.

28. Aristotle, *On Prophesying by Dreams,* p. 628. Aristotle, *Parts of Animals,* pp. 656–57. Thales, quoted by Diogenes Laertius, *Lives of Eminent Philosophers* (Cambridge: Harvard University Press, 1925), I, p. 37.

29. Aristotle, *Metaphysics,* p. 692.

30. Aristotle, *On The Soul,* p. 538.

31. Aristotle, *Nicomachean Ethics,* p. 944.

32. Heisenberg, *Physics and Philosophy,* p. 180.

33. Schrödinger, *Science and Humanism,* p. 57.

34. Wigner, *Symmetries and Reflections,* pp. 176–77.

35. Mozart, quoted by Machlis, *The Enjoyment of Music,* p. 308.

36. Jean August Ingres, in *Artists on Art,* p. 218.

37. Stravinsky, in *Composers on Music,* p. 452.

38. T.S. Eliot, "Tradition and the Individual Talent," in *The Sacred Wood* (London: Methuen, 1920), pp. 44–49.

VIII   THE PRESENT

1. Schrödinger, *Mind and Matter,* pp. 41, 66, 68.

2. Sperry, "Interview," p. 74.

3. Schrödinger, *Nature and the Greeks* (Cambridge: Cambridge University Press, 1954), p. 93.

4. Heisenberg, *Physics and Philosophy,* pp. 197–98.

5. Darwin, *The Origin of Species and The Descent of Man,* p. 914.

6. Heisenberg, "Unity of the Scientific Outlook," in *Philosophical Problems,* p. 105.

7. Pp. 92–93.

8. Darwin, p. 148.

9. Eccles, *Facing Reality,* pp. 171, 4. Joseph Weizenbaum, *Computer Power and Human Reason: From Judgment to Calculation* (San Francisco: W.H. Freeman, 1976), p. 203.

10. Eccles, *The Human Mystery,* p. vii.

11. S.P. Thompson, *The Life of Lord Kelvin* (New York: Chelsea, 1977), II, p. 835.

12. Einstein and Infeld, *The Evolution of Physics,* p. 121.

13. Bacon, *The New Organon,* pp. 77, 122, 53.

14. Pp. 8–9.

15. Galileo, quoted by Burtt, p. 77.

16. Galileo, *Dialogue on the Great World Systems,* trans. Thomas Salusbury (Chicago: University of Chicago Press, 1953), pp. 59–60.

17. Descartes, letter to Mersenne, 18 December 1629, quoted in *Towards a Mechanistic Philosophy* (Keynes, England: Open University Press, 1974), p. 21. Blaise Pascal, *Provincial Letters, Pensees, Scientific Treatises,* Vol. 33 of *Great Books of the Western World* (Chicago: Encyclopedia Britannica, 1952), p. 356.

18. Bacon, p. 29.

19. P. 8.

20. Descartes, "Discourse on Method," pp. 130–31.

21. Infeld, pp. 18–19.

22. Antoine Lavoisier, *Elements of Chemistry* (Ann Arbor: Edward Brothers, 1945), Preface.

23. Gunnar Myrdal, *An American Dilemma* (New York: Harper & Row, 1949), p. 92.

24. Yuri Glazov, "The Soviet Intelligentsia, Dissents and the West," *Studies in Soviet Thought,* (1979), pp. 327–28.

25. Wigner, *Symmetries and Reflections,* p. 172.

BIBLIOGRAPHY

Adams, Hazard. *Critical Theory Since Plato*. New York: Harcourt Brace, 1971.

Allen, Walter, ed., *Writers on Writing*. New York: Dutton, 1949.

Aquinas, Thomas. *Summa Theologica*. New York: Benziger Brothers, 1947.

———. *The Pocket Aquinas*. Ed. Vernon J. Bourke. New York: Pocket Books, 1960.

Aristotle. *The Basic Works of Aristotle*. Ed. Richard McKeon. New York: Random House, 1941.

Augustine. *The City of God*. Trans. G. Walsh, *et al*. Garden City: Doubleday, 1950.

Bacon, Francis. *The New Organon and Related Writings*. Ed. Fulton H. Anderson. Indianapolis: Bobbs-Merrill, 1960.

Bath, Geoffrey, *The State of the Universe*. Oxford: Clarendon Press, 1980.

Bell, Eric Temple. *Men of Mathematics*. New York: Simon & Schuster, 1937.

Bentley, W.A. *Snow Crystals*. New York: Dover, 1962.

Berkeley, George. *A Treatise Concerning The Principles of Human Knowledge*. Indianapolis: Bobbs-Merrill, 1957.

_____. "Philosophical Commentaries, No. 24" in *The Works of George Berkeley*. 9 Vols. Ed. A.A. Luce and T.E. Jessop. London: Thomas Nelson, 1948.

Berry, Thomas. *The New Story*. Chambersburg, Pa.: Anima Publications, 1978.

Bludman, Sidney A. "Thermodynamics and the end of a closed Universe." *Nature* 308 (22 March 1984), pp. 319–322.

Bohr, Niels. *Essays, 1958–1962, On Atomic Physics and Human Knowledge*. New York: Interscience, 1963.

Born, Max. *Physics in My Generation*. London & New York: Pergamon Press, 1956.
_____. *The Restless Universe*. New York: Dover, 1951.

Browning, Elizabeth Barrett. *The Complete Works of Elizabeth Barrett Browning*. Ed. Charlotte Porter and Helen A. Clarke. New York: Crowell, 1900.

Burtt, Edwin Arthur. *The Metaphysical Foundations of Modern Physical Science*. 2nd ed. 1932. Rpt. Garden City: Doubleday, 1954.

Camus, Albert. *The Myth of Sisyphus & Other Essays*. Trans. Justin O'Brien. New York: Vintage Books, 1955.

Carr. B.J. and M.J. Rees. "The Anthropic Principle and the Structure of the Physical World." *Nature* 278 (12 April 1979), pp. 605–612.

Carter, Brandon. "Large Number Coincidences and the Anthropic Principle." In *Confrontation of Cosmological Theories with Observational Data,* ed. M.S. Longair. Dordrecht, Holland: Reidel, 1974.

Chandrasekhar, S. "Beauty and the Quest for Beauty in Science." *Physics Today* 32 (1979), 25–30.

Child, Irvin L. *Humanistic Psychology and the Research Tradition: Their Several Virtues.* New York: John Wiley, 1973.

Cicero, *De Natura Deorum.* Trans. H. Rackham. Cambridge: Harvard University Press, 1933.

Clifford, William Kingdon. "Mind and Body." In *Lectures and Essays.* Ed. Leslie Stephen and F. Pollack. London: Macmillian, 1979.

Copland, Aaron. *What to Listen for in Music.* New York: McGraw-Hill, 1957.

Darwin, Charles. *The Origin of Species and The Descent of Man.* New York: Modern Library, n.d.

de Broglie, Louis. *Savants et Découvertes.* Paris: Albin Michel, 1951.

_____. *La Physique Nouvelle et les Quanta.* Paris: Flammarion, 1937.

Descartes, René. *Méditations* in *Descartes Oeuvres et Lettres.* Ed. André Bridoux. Paris: Editions Gallimard, 1953.

_____. *Philosophical Writings.* Trans. Norman K. Smith. New York: Modern Library, 1958.

Diogenes Laertius. *Lives of Eminent Phi-

losophers. Trans. R.D. Hicks. 2 vols. Cambridge, Mass.: Harvard University Press, 1925.

Dirac, P.A.M. "The Evolution of the Physicist's Picture of Nature." *Scientific American* 208 (May 1963), 45–53.

Dyson, Freeman. *Disturbing the Universe.* New York: Harper & Row, 1979.

Eccles, John C. *Facing Reality: Philosophical Adventures by a Brain Scientist.* Berlin and New York: Springer-Verlag, 1970.

──────. *The Human Mystery.* New York: Springer-Verlag, 1979.

Einstein, Albert. *Out of My Later Years.* New York: Philosophical Library, 1950.

──────. "Autobiographical Notes." In *Albert Einstein: Philosopher-Scientist,* ed. Paul Schilpp. New York: Harper & Row, 1959.

──────, and Leopold Infeld. *The Evolution of Physics.* New York: Simon & Schuster, 1966.

Eliot, T.S. *The Sacred Wood.* London: Methuen, 1920.

Epictetus. "The Discourses." In *The Works of Epictetus.* Trans. Thomas Higginson. New York: Nelson, 1890.

Feynman, Richard. *The Character of Physical Law.* Cambridge, Mass.: M.I.T. Press, 1965.

──────, and Murray Gell-Mann. *Physical Review* 109 (1958), 193–197.

Frankl, Viktor E. *Man's Search for Meaning*. New York: Pocket Books, 1959.

_____. *The Will to Meaning*. New York: New American Library, 1969.

_____. *The Doctor and the Soul*. New York: Knopf, 1955.

_____. *The Unconscious God*. New York: Simon & Schuster, 1975.

_____. *Psychoanalysis and Existentialism*. New York: Simon & Schuster, 1967.

Freud, Sigmund. *Civilization and Its Discontents*. Trans. James Strachey. New York: Norton, 1962.

_____. *The Future of an Illusion*. Trans. W.D. Robson-Scott. Garden City: Doubleday, 1964.

_____. "Psychoanalysis and Telepathy." In *Standard Edition of the Complete Psychological Works of Sigmund Freud*, Vol. 18. Ed. James Strachey. London: Hogarth Press, 1955.

Galileo. *Dialogue on the Great World Systems*. Trans. Thomas Salusbury. Chicago: University of Chicago Press, 1953.

_____. *Discoveries and Opinions of Galileo*. Trans. Stillman Drake. Garden City: Doubleday, 1957.

Glazov, Yuri. "The Soviet Intelligentsia, Dissents and the West." *Studies in Soviet Thought*, 19 (1979), 321–344.

Goldwater, Robert, and Marco Treves, eds.

Artists on Art. New York: Pantheon, 1947.

Hatterer, Lawrence John. "Work Identity: A New Psychotherapeutic Dimension." *Psychiatric Spectator,* 2, No. 7, (1965).

Hawking, Steven W. "The Anisotropy of the Universe at Large Times." In *Confrontation of Cosmological Theories with Observational Data,* ed. M.S. Longair. Dordrecht, Holland: Reidel, 1974.

Heisenberg, Werner. *Philosophical Problems of Nuclear Science.* New York: Pantheon Books, 1952.

———. *Physics and Philosophy.* New York: Harper & Row, 1958.

———. *Across the Frontier.* New York: Harper & Row, 1974.

———. *Physics and Beyond.* New York: Harper & Row, 1972.

Hobbes, Thomas. *Leviathan.* Indianapolis: Bobbs-Merrill, 1958.

Hume, David. *A Treatise of Human Nature.* Oxford: Clarendon Press, 1888.

Huxley, Thomas. *Selections From the Essays of T.H. Huxley.* Ed. Alburey Castell. New York: Appleton-Century-Crofts, 1948.

Infeld, Leopold. *Albert Einstein: His Work and Its Influence on Our World.* New York: Scribner's, 1950.

Jastrow, Robert. *God and the Astronomers.* New York: Norton, 1978.

Judson, Horace Freeland. *The Search for*

Solutions. New York: Holt, Rinehart & Winston, 1980.

Kant, Immanuel. *Prolegomena to Any Future Metaphysics*. Indianapolis: Bobbs-Merrill, 1950.

Kaufmann, William. *The Cosmic Frontiers of General Relativity*. Boston: Little, Brown, 1977.

Kline, Morris. *Mathematics: The Loss of Certainty*. New York: Oxford University Press, 1980.

Lavoisier, Antoine. *Elements of Chemistry*. Ann Arbor: Edward Brothers, 1945.

Mach, Ernst. "On Instruction in the Classics and the Sciences." In *Popular Scientific Lectures,* trans. Thomas J. McCormack. La Salle, Illinois: Open Court, 1943.

Machlis, Joseph. *The Enjoyment of Music*. New York: Norton, 1963.

Mann, Thomas. *Death in Venice and Seven Other Stories*. Trans. H.T. Lowe-Porter. New York: Vintage, 1954.

Margenau, Henry. *The Miracle of Existence*. Woodbridge, Connecticut: Ox Bow Press, 1984.

Maslow, Abraham H. *Toward A Psychology of Being*. New York: Van Nostrand Reinhold, 1968.

May, Rollo. *Psychology and the Human Dilemma*. New York: Van Nostrand, 1967.

Morgenstern, Sam, ed. *Composers on Music*. New York: Bonanza, 1956.

Morowitz, Harold. *The Wine of Life and Other Essays on Societies, Energy and Living Things.* New York: St. Martin's, 1979.

Myrdal, Gunnar. *An American Dilemma.* New York: Harper & Row, 1944.

Newton, Isaac. *Principia.* Trans. Florian Cajori. Berkeley: University of California Press, 1934.

––––––. *Opticks.* New York: Dover, 1952.

––––––. *Newton's Philosophy of Nature: Selections From His Writings,* ed. H.S. Thayer. New York: Macmillan, 1953.

Nietzsche, Friedrich. *Beyond Good and Evil.* Trans. Marianne Cowan. Chicago: Henry Regnery, 1955.

Pascal, Blaise. *Provincial Letters, Pensees, Scientific Treatises.* Vol. 33 of *Great Books of the Western World.* Chicago: Encyclopedia Britannica, 1952.

Penfield, Wilder. *The Mystery of the Mind: A Critical Study of Consciousness and the Human Brain.* Princeton: Princeton University Press, 1975.

Poincaré, Henri. *Science et Méthode.* Paris: Flammarion, 1949.

––––––. *The Value of Science.* New York: Dover, 1958.

Portmann, Adolf. *Animal Forms and Patterns.* New York: Schocken, 1967.

––––––. *New Paths in Biology.* New York: Harper & Row, 1964.

Rogers, Carl, R. "Toward a Science of the Person." In *Behaviorism and*

Phenomenology: Contrasting Bases for Modern Psychology, ed. T.W. Wann. Chicago: University of Chicago Press, 1964.

Russell, Bertrand. *Why I Am Not A Christian.* New York: Simon & Schuster, 1957.

Sartre, Jean-Paul. *Existentialism and Human Emotions.* Trans. Bernard Frechtman. New York: Citadel Press, 1957.

Schneider, Pierre. *The World of Manet.* New York: Time-Life Books, 1968.

Schrödinger, Erwin. *Mind and Matter.* Cambridge: Cambridge University Press, 1958.

_____. *Nature and the Greeks.* Cambridge: Cambridge University Press, 1954.

_____. *Science and Humanism.* Cambridge: Cambridge University Press, 1951.

Severin, Frank T. *Discovering Man in Psychology: A Humanistic Approach.* New York: McGraw-Hall, 1973.

Sherrington, Charles. *Man on His Nature.* Cambridge: Cambridge University Press, 1975.

Silk, Joseph. *The Big Bang: The Creation and Evolution of the Universe.* San Francisco: W.H. Freeman, 1980.

Snow, C.P. *Public Affairs.* New York: Scribner's, 1971.

Solzhenitsyn, Aleksandr. *The Gulag*

Archipelago Two. Trans. Thomas P. Whitney. New York: Harper & Row, 1975.

Sorell, Walter. *The Duality of Vision*. Indianapolis: Bobbs-Merrill, 1970.

Sperry, Roger. "Mind, Brain and Humanist Values." In *New Views of the Nature of Man,* ed. John R. Platt. Chicago: University of Chicago Press, 1965.

————. "Interview." *Omni,* August 1983.

Tartarkiewicz, Wladyslaw. *History of Aesthetics*. Ed. D. Petsch. The Hague: Mouton, 1974.

Taubes, Gary. "The Snowflake Enigma." *Discover,* Jan. 1984, 75–78.

Taylor, Gordon Rattray. *The Natural History of Mind*. New York: Dutton, 1979.

Thompson, S.P. *The Life of Lord Kelvin*. 2nd ed. 2 vols. New York: Chelsea, 1977.

Thomson, Sir George. *The Inspiration of Science*. Oxford: Oxford University Press, 1961.

Thoreau, Henry David. *Thoreau on Man and Nature*. Ed. Arthur Volkman. Mount Vernon, New York: Peter Pauper Press, 1960.

————. *Journal*. Selections in *The Wilderness Is the Preservation of the World*. Ed. Eliot Porter. New York: Ballantine Books, 1967.

Waddington, C.H. *Towards a Theoretical Biology*. Chicago: Aldine, 1969.

Watson, James. *The Double Helix*. New York: Mentor, 1968.

Watson, John B. *Behaviorism*. Chicago: University of Chicago Press, 1930.

_____. *Psychology From the Standpoint of a Behaviorist*. Philadelphia: Lippincott, 1919.

Watts, Alan W. *The Way of Zen*. New York: Random House, 1957.

Weinberg, Steven. *The First Three Minutes*. New York: Basic Books, 1977.

_____. "Science's 'Parallels with Art'—a Physicist's View." *U.S. News & World Report,* 8 Sept. 1980, 68.

Weizenbaum, Joseph. *Computer Power and Human Reason: From Judgment to Calculation*. San Francisco: W.H. Freeman, 1976.

Weizacker, Carl von. *The World View of Physics*. Chicago: University of Chicago Press, 1952.

Wheeler, John A. "Genesis and Observership." In *Foundational Problems in the Special Sciences,* ed. Robert E. Butts and Jaakko Hintikka. Dordrecht, Holland: Reidel, 1977.

_____. "The Universe as a Home for Man." *American Scientist* 62 (Nov.-Dec. 1974), 683–691.

Wigner, Eugene. *Symmetries and Reflections*. Bloomington: Indiana University Press, 1967.

INDEX

A

P